Flinn Scientific
ChemTopic™ Labs

Chemistry of Organic Compounds

Senior Editor

Irene Cesa
Flinn Scientific, Inc.
Batavia, IL

Curriculum Advisory Board

Bob Becker
Kirkwood High School
Kirkwood, MO

Kathleen J. Dombrink
McCluer North High School
Florissant, MO

Robert Lewis
Downers Grove North High School
Downers Grove, IL

John G. Little
St. Mary's High School
Stockton, CA

Lee Marek
University of Illinois–Chicago
Chicago, IL

John Mauch
Braintree High School
Braintree, MA

Dave Tanis
Grand Valley State University
Allendale, MI

FLINN SCIENTIFIC INC.
"Your Safer Source for Science Supplies"
P.O. Box 219 • Batavia, IL 60510
1-800-452-1261 • www.flinnsci.com

ISBN 978-1-877991-89-9

Copyright © 2006 Flinn Scientific, Inc.

All rights reserved. No part of this book may be reproduced or transmitted in any form or by any means, electronic or mechanical, including, but not limited to photocopy, recording, or any information storage and retrieval system, without permission in writing from Flinn Scientific, Inc.
No part of this book may be included on any Web site.

Reproduction permission is granted only to the science teacher who has purchased this volume of Flinn ChemTopic™ Labs, Chemistry of Organic Compounds, Catalog No. AP6987 from Flinn Scientific, Inc. Science teachers may make copies of the reproducible student pages for use only by their students.

Printed in the United States of America.

Table of Contents

	Page
Flinn ChemTopic™ Labs Series Preface	i
About the Curriculum Advisory Board	ii
Chemistry of Organic Compounds Preface	iii
Format and Features	iv–v
Experiment Summaries and Concepts	vi–vii

Introduction to Organic Chemistry

Introduction to Functional Groups	1
Naming Organic Compounds	3

Experiments

Models of Organic Compounds	5
Making Soap	17
Preparation of Esters	29
Synthesis of Aspirin	41
Steam Distillation of Cinnamon	53

Demonstrations

Cleaning with Charcoal	65
The Carbon Soufflé	67
Feeling Blue	69
Kaleidoscope . . . Optical Activity	71
Salt-Out the Red, White and Blue	75

Supplementary Information

Safety and Disposal Guidelines	78
National Science Education Standards	80
Master Materials Guide	82

Flinn ChemTopic™ Labs Series Preface
Lab Manuals Organized Around Key Content Areas in Chemistry

In conversations with chemistry teachers across the country, we have heard a common concern. Teachers are frustrated with their current lab manuals, with experiments that are poorly designed and don't teach core concepts, with procedures that are rigid and inflexible and don't work. Teachers want greater flexibility in their choice of lab activities. As we further listened to experienced master teachers who regularly lead workshops and training seminars, another theme emerged. Master teachers mostly rely on collections of experiments and demonstrations they have put together themselves over the years. Some activities have been passed on like cherished family recipe cards from one teacher to another. Others have been adapted from one format to another to take advantage of new trends in microscale equipment and procedures, technology innovations, and discovery-based learning theory. In all cases the experiments and demonstrations have been fine-tuned based on real classroom experience.

Flinn Scientific has developed a series of lab manuals based on these "cherished recipe cards" of master teachers with proven excellence in both teaching students and training teachers. Created under the direction of an Advisory Board of award-winning chemistry teachers, each lab manual in the Flinn ChemTopic™ Labs series contains 4–6 student-tested experiments that focus on essential concepts and applications in a single content area. Each lab manual also contains 4–6 demonstrations that can be used to illustrate a chemical property, reaction, or relationship and will capture your students' attention. The experiments and demonstrations in the Flinn ChemTopic™ Labs series are enjoyable, highly focused, and will give students a real sense of accomplishment.

Laboratory experiments allow students to experience chemistry by doing chemistry. Experiments have been selected to provide students with a crystal-clear understanding of chemistry concepts and encourage students to think about these concepts critically and analytically. Well-written procedures are guaranteed to work. Reproducible data tables teach students how to organize their data so it is easily analyzed. Comprehensive teacher notes include a master materials list, solution preparation guide, complete sample data, and answers to all questions. Detailed lab hints and teaching tips show you how to conduct the experiment in your lab setting and how to identify student errors and misconceptions before students are led astray.

Chemical demonstrations provide another teaching tool for seeing chemistry in action. Because they are both visual and interactive, demonstrations allow teachers to take students on a journey of observation and understanding. Demonstrations provide additional resources to develop central themes and to magnify the power of observation in the classroom. Demonstrations using discrepant events challenge student misconceptions that must be broken down before new concepts can be learned. Use demonstrations to introduce new ideas, illustrate abstract concepts that cannot be covered in lab experiments, and provide a spark of excitement that will capture student interest and attention.

Safety, flexibility, and choice

Safety always comes first. Depend on Flinn Scientific to give you upfront advice and guidance on all safety and disposal issues. Each activity begins with a description of the hazards involved and the necessary safety precautions to avoid exposure to these hazards. Additional safety, handling, and disposal information is also contained in the teacher notes.

The selection of experiments and demonstrations in each Flinn ChemTopic™ Labs manual gives you the flexibility to choose activities that match the concepts your students need to learn. No single teacher will do all of the experiments and demonstrations with a single class. Some experiments and demonstrations may be more helpful with a beginning-level class, while others may be more suitable with an honors class. All of the experiments and demonstrations have been keyed to national content standards in science education.

Chemistry is an experimental science!

Whether they are practicing key measurement skills or searching for trends in the chemical properties of substances, all students will benefit from the opportunity to discover chemistry by doing chemistry. No matter what chemistry textbook you use in the classroom, Flinn ChemTopic™ Labs will help you give your students the necessary knowledge, skills, attitudes, and values to be successful in chemistry.

About the Curriculum Advisory Board

Flinn Scientific is honored to work with an outstanding group of dedicated chemistry teachers. The members of the Flinn ChemTopic Labs Advisory Board have generously contributed their proven experiments and demonstrations to create these topic lab manuals. The wisdom, experience, creativity, and insight reflected in their lab activities guarantee that students who perform them will be more successful in learning chemistry. On behalf of all chemistry teachers, we thank the Advisory Board members for their service and dedication to chemistry education.

Bob Becker teaches chemistry and AP chemistry at Kirkwood High School in Kirkwood, MO. Bob received his B.A. from Yale University and M.Ed. from Washington University and has 20 years of teaching experience. A well-known demonstrator, Bob has conducted more than 100 demonstration workshops across the U.S. and Canada and was a Team Leader for the Flinn Foundation Summer Workshop Program. His creative and unusual demonstrations have been published in the *Journal of Chemical Education,* the *Science Teacher,* and *Chem13 News*. Bob is the author of two books of chemical demonstrations, *Twenty Demonstrations Guaranteed to Knock Your Socks Off, Volumes I and II,* published by Flinn Scientific. Bob has been awarded the James Bryant Conant Award in High School Teaching from the American Chemical Society, the Regional Catalyst Award from the Chemical Manufacturers Association, and the Tandy Technology Scholar Award.

Kathleen J. Dombrink teaches chemistry and advanced-credit college chemistry at McCluer North High School in Florissant, MO. Kathleen received her B.A. in Chemistry from Holy Names College and M.S. in Chemistry from St. Louis University and has 35 years of teaching experience. Recognized for her strong support of professional development, Kathleen has been selected to participate in the Fulbright Memorial Fund Teacher Program in Japan and NEWMAST and Dow/NSTA Workshops. She served as co-editor of the inaugural issues of *Chem Matters* and was a Woodrow Wilson National Fellowship Foundation Chemistry Team Member for 11 years. Kathleen is currently a Team Leader for the Flinn Foundation Summer Workshop Program. Kathleen has received the Presidential Award, the Midwest Regional Teaching Award from the American Chemical Society, the Tandy Technology Scholar Award, and a Regional Catalyst Award from the Chemical Manufacturers Association.

Robert Lewis retired from teaching chemistry at Downers Grove North High School in Downers Grove, IL, and is currently a Secondary Coordinator for the GATE program in Chicago. Robert received his B.A. from North Central College and M.A. from University of the South and has 30 years of teaching experience. He was a founding member of Weird Science, a group of chemistry teachers that traveled throughout the country to stimulate teacher enthusiasm for using demonstrations to teach science. Robert served as a Team Leader for both the Woodrow Wilson National Fellowship Foundation and the Flinn Foundation Summer Workshop Program. Robert has received the Presidential Award, the James Bryant Conant Award in High School Teaching from the American Chemical Society, the Tandy Technology Scholar Award, a Regional Catalyst Award from the Chemical Manufacturers Association, and a Golden Apple Award from the State of Illinois.

John G. Little teaches chemistry and AP chemistry at St. Mary's High School in Stockton, CA. John received his B.S. and M.S. in Chemistry from University of the Pacific and has 39 years of teaching experience. Highly respected for his well-designed labs, John is the author of two lab manuals, *Chemistry Microscale Laboratory Manual* (D. C. Heath), and *Microscale Experiments for General Chemistry* (with Kenneth Williamson, Houghton Mifflin). He is also a contributing author to *Science Explorer* (Prentice Hall) and *World of Chemistry* (McDougal Littell). John served as a Chemistry Team Leader for both the Woodrow Wilson National Fellowship Foundation and the Flinn Foundation Summer Workshop Program. He has been recognized for his dedicated teaching with the Tandy Technology Scholar Award and the Regional Catalyst Award from the Chemical Manufacturers Association.

Lee Marek retired from teaching chemistry at Naperville North High School in Naperville, IL and currently teaches at the University of Illinois–Chicago. Lee received his B.S. in Chemical Engineering from the University of Illinois and M.S. degrees in Physics and Chemistry from Roosevelt University. He has more than 30 years of teaching experience and is a Team Leader for the Flinn Foundation Summer Workshop Program. His students have won national recognition in the International Chemistry Olympiad, the Westinghouse Science Talent Search, and the Internet Science and Technology Fair. Lee was a founding member of Weird Science and has presented more than 500 demonstration and teaching workshops for more than 300,000 students and teachers across the country. Lee has performed science demonstrations on the *David Letterman Show* 20 times. Lee has received the Presidential Award, the James Bryant Conant Award in High School Teaching and the Helen M. Free Award for Public Outreach from the American Chemical Society, the National Catalyst Award from the Chemical Manufacturers Association, and the Tandy Technology Scholar Award.

John Mauch teaches chemistry and AP chemistry at Braintree High School in Braintree, MA. John received his B.A. in Chemistry from Whitworth College and M.A. in Curriculum and Education from Washington State University and has more than 25 years of teaching experience. John is an expert in microscale chemistry and is the author of two lab manuals, *Chemistry in Microscale, Volumes I and II* (Kendall/Hunt). He is also a dynamic and prolific demonstrator and workshop leader. John has presented the Flinn Scientific Chem Demo Extravaganza show at NSTA conventions for eight years and has conducted more than 100 workshops across the country. John was a Chemistry Team Member for the Woodrow Wilson National Fellowship Foundation program and is currently a Board Member for the Flinn Foundation Summer Workshop Program. John has received the Massachusetts Chemistry Teacher of the Year Award from the New England Institute of Chemists.

Dave Tanis is Associate Professor of Chemistry at Grand Valley State University in Allendale, MI. Dave received his B.S. in Physics and Mathematics from Calvin College and M.S. in Chemistry from Case Western Reserve University. He taught high school chemistry for 26 years before joining the staff at Grand Valley State University to direct a coalition for improving pre-college math and science education. Dave later joined the faculty at Grand Valley State University and currently teaches courses for pre-service teachers. The author of two laboratory manuals, Dave acknowledges the influence of early encounters with Hubert Alyea, Marge Gardner, Henry Heikkinen, and Bassam Shakhashiri in stimulating his long-standing interest in chemical demonstrations and experiments. Continuing this tradition of mentorship, Dave has led more than 40 one-week institutes for chemistry teachers and served as a Team Member for the Woodrow Wilson National Fellowship Foundation for 13 years. He is currently a Board Member for the Flinn Foundation Summer Workshop Program. Dave received the College Science Teacher of the Year Award from the Michigan Science Teachers Association.

Preface
Chemistry of Organic Compounds

Attaching the word *organic* to food and other consumer items has become a mark of quality, to signify that something is natural and unadulterated. Isn't it time to reclaim that connection for organic chemistry as well? Organic chemistry is the chemistry of life! The science of organic chemistry has its roots in the study of natural products—organic compounds isolated from nature. Organic compounds are foods and medicines, soaps and perfumes, preservatives and cosmetics, spices and seasonings, etc. The purpose of *Chemistry of Organic Compounds,* Volume 19 in the Flinn ChemTopic™ Labs series, is to provide high school chemistry teachers with meaningful, easy-to-do laboratory activities that will help students apply the principles of general chemistry to organic compounds. Five experiments and five demonstrations allow students to understand the central role of organic compounds in a wide range of consumer goods and applications. All activities include a discussion of historical perspectives that will make the stories of these fascinating compounds come alive for students.

Introduction to Organic Chemistry

More than nine million organic compounds are known! How are they similar? How are they different? Two introductory worksheets, "Introduction to Functional Groups" and "Naming Organic Compounds," will give students a basic understanding of the structure of organic compounds and will help students recognize relationships among organic compounds. In "Models of Organic Compounds," a guided-inquiry activity, students build organic molecules from the ground up using models and use the reasoning skills of the scientific method to draw structural formulas, determine the general formulas of different classes of compounds, and develop the concept of isomers. There are also several good lead-in demonstrations to illustrate the general properties of organic compounds. Compare and contrast the miscibility and density of organic solvents with water in "Salt-Out the Red, White and Blue," or use the classic "Feeling Blue" demonstration to apply redox principles to organic chemistry.

Natural Products and Consumer Science

The transformation of organic compounds from natural products into consumer products has been a dominant theme in the history of organic chemistry. Having students investigate the preparation and properties of organic consumer products in the lab helps to reinforce this important theme. In "Making Soap—The Oldest Organic Reaction," students make homemade bars of soap using natural fats and oils and study the properties of soap. Use the ancient craft of soap-making to show students what chemists really do—chemists use chemicals to make things people use every day! Continue the fascinating trail of discovery from natural products to consumer products with "Preparation of Esters" and "Synthesis of Aspirin." The mini-scale preparations of these organic compounds provide an important safety benefit and save valuable time in the high school lab. The ester lab, which is written as a cooperative class project, highlights the wide variety of fragrant esters in nature and teaches students that organic chemistry makes good "scents"! In "Steam Distillation of Cinnamon," students are introduced to distillation and extraction, common laboratory techniques that have been used for centuries to obtain essential oils and other natural products.

Science in Personal and Social Perspectives

Building connections between the sciences and integrating social and personal perspectives are important goals of science education, and indeed these goals are formally embedded in the National Science Education Standards. The activities in *Chemistry of Organic Compounds* offer an excellent opportunity to help students see chemistry not just in a textbook, but in the world around them. All of the experiments and demonstrations have been optimized to adapt them to the knowledge and skill level of the high school chemistry curriculum. The use of hazardous reagents has been critically evaluated, preparations have been scaled down, and procedures have been reviewed and simplified to make sure they are as safe as possible yet still produce satisfying outcomes. All of the activities have been thoroughly tested and retested. You know they will work! Use the experiment summaries and concepts on the following pages to locate the concepts you want to teach and to choose experiments and demonstrations that will help you meet your goals.

Format and Features

Flinn ChemTopic™ Labs

All experiments and demonstrations in Flinn ChemTopic™ Labs are printed in a 10⅞″ × 11″ format with a wide 2″ margin on the inside of each page. This reduces the printed area of each page to a standard 8½″ × 11″ format suitable for copying.

The wide margin assures you the entire printed area can be easily reproduced without damaging the binding. The margin also provides a convenient place for teachers to add their own notes.

Concepts — Use these bulleted lists along with state and local standards, lesson plans, and your textbook to identify activities that will allow you to accomplish specific learning goals and objectives.

Background — A balanced source of information for students to understand why they are doing an experiment, what they are doing, and the types of questions the activity is designed to answer. This section is not meant to be exhaustive or to replace the students' textbooks, but rather to identify the core concepts that should be covered before starting the lab.

Experiment Overview — Clearly defines the purpose of each experiment and how students will achieve this goal. Performing an experiment without a purpose is like getting travel directions without knowing your destination. It doesn't work, especially if you run into a roadblock and need to take a detour!

Pre-Lab Questions — Making sure that students are prepared for lab is the single most important element of lab safety. Pre-lab questions introduce new ideas or concepts, review key calculations, and reinforce safety recommendations. The pre-lab questions may be assigned as homework in preparation for lab or they may be used as the basis of a cooperative class activity before lab.

Materials — Lists chemical names, formulas, and amounts for all reagents—along with specific glassware and equipment—needed to perform the experiment as written. The material dispensing area is a main source of student delay, congestion, and accidents. Three dispensing stations per room are optimum for a class of 24 students working in pairs. To safely substitute different items for any of the recommended materials, refer to the *Lab Hints* section in each experiment or demonstration.

Safety Precautions — Instruct and warn students of the hazards associated with the materials or procedure and give specific recommendations and precautions to protect students from these hazards. Please review this section with students before beginning each experiment.

Procedure — This section contains a stepwise, easy-to-follow procedure, where each step generally refers to one action item. Contains reminders about safety and recording data where appropriate. For inquiry-based experiments the procedure may restate the experiment objective and give general guidelines for accomplishing this goal.

Data Tables — Data tables are included for each experiment and are referred to in the procedure. These are provided for convenience and to teach students the importance of keeping their data organized in order to analyze it. To encourage more student involvement, many teachers prefer to have students prepare their own data tables. This is an excellent pre-lab preparation activity—it ensures that students have read the procedure and are prepared for lab.

Post-Lab Questions or Data Analysis — This section takes students step-by-step through what they did, what they observed, and what it means. Meaningful questions encourage analysis and promote critical thinking skills. Where students need to perform calculations or graph data to analyze the results, these steps are also laid out sequentially.

Format and Features
Teacher's Notes

Master Materials List — Lists the chemicals, glassware, and equipment needed to perform the experiment. All amounts have been calculated for a class of 30 students working in pairs. For smaller or larger class sizes or different working group sizes, please adjust the amounts proportionately.

Preparation of Solutions — Calculations and procedures are given for preparing all solutions, based on a class size of 30 students working in pairs. With the exception of particularly hazardous materials, the solution amounts generally include 10% extra to account for spillage and waste. Solution volumes may be rounded to convenient glassware sizes (100-mL, 250-mL, 500-mL, etc.).

Safety Precautions — Repeats the safety precautions given to the students and includes more detailed information relating to safety and handling of chemicals and glassware. Refers to Material Safety Data Sheets that should be available for all chemicals used in the laboratory.

Disposal — Refers to the current *Flinn Scientific Catalog/Reference Manual* for general guidelines and specific procedures governing the disposal of laboratory waste. Because we recommend that teachers review local regulations before beginning any disposal procedure, the information given in this section is for general reference purposes only. However, if a disposal step is included as part of the experimental procedure itself, then the specific solutions needed for disposal are described in this section.

Lab Hints — This section reveals common sources of student errors and misconceptions and where students are likely to need help. Identifies the recommended length of time needed to perform each experiment, suggests alternative chemicals and equipment that may be used, and reminds teachers about new techniques (filtration, pipeting, etc.) that should be reviewed prior to lab.

Teaching Tips — This section puts the experiment in perspective so that teachers can judge in more detail how and where a particular experiment will fit into their curriculum. Identifies the working assumptions about what students need to know in order to perform the experiment and answer the questions. Highlights historical background and applications-oriented information that may be of interest to students.

Sample Data — Complete, actual sample data obtained by performing the experiment exactly as written is included for each experiment. Student data will vary.

Answers to All Questions — Representative or typical answers to all questions. Includes sample calculations and graphs for all data analysis questions. Information of special interest to teachers only in this section is identified by the heading "Note to the teacher." Student answers will vary.

Look for these icons in the *Experiment Summaries and Concepts* section and in the *Teacher's Notes* of individual experiments to identify inquiry-, microscale-, and technology-based experiments, respectively.

Experiment Summaries and Concepts

Experiment

Models of Organic Compounds—Guided Inquiry

There are more than nine million organic compounds! What makes all these compounds different? Introduce the basic structural theory of organic chemistry by building molecules from the ground up using models. Students will actually use the reasoning skills of the scientific method as they follow this guided-inquiry activity to draw structural formulas of organic molecules, determine the general formulas of different classes of compounds, and develop the concept of isomers. Lecture is no substitute for holding models in your hands, rotating them and turning them upside down, or taking them apart and putting them together again a different way!

Making Soap—The Oldest Organic Reaction

Soap-making is an ancient craft and one of the oldest known chemical reactions involving organic compounds. It is also one of the best ways to show students what chemists really do. Chemists don't just mix chemicals to make new chemicals—they use chemicals to make things that people use everyday! All students, no matter how laid-back or jaded, will be amazed and proud when they hold up a bar of soap they made themselves in the lab!

Preparation of Esters—Nature's Flavors and Fragrances

What do you taste when you bite into an apple or a banana? The first "taste" of any food is actually the aroma or fragrance of volatile organic compounds. In the case of fruits, the "organic" flavor is due to esters. Learn about the structure, preparation, and properties of esters with this great cooperative class project. With five alcohols and four carboxylic acids to choose from, the entire class can synthesize a wide variety of common esters that are used as fragrance and flavor additives. Discover why organic chemistry makes good "scents"!

Synthesis of Aspirin—From Natural Products to Painkillers

Aspirin, first synthesized in 1897, is one of the oldest yet most common drugs in use today. This one-time "wonder drug" is routinely prescribed today to prevent heart attacks and strokes, especially among the elderly. Students trace the path of discovery for aspirin from a natural folk remedy that was used for thousands of years to the first modern synthetic designer drug! The purpose of this experiment is to prepare aspirin (acetylsalicylic acid), determine its purity, and investigate its chemical properties.

Steam Distillation of Cinnamon—Cinnamaldehyde and Oil of Cinnamon

Looking for a way to introduce students to organic chemistry? Take them back to the roots of the science—the study of natural products. Steam distillation is the most common method for isolating essential oils, natural products that have been used since ancient times as perfumes, flavorings, and even medicines. Students obtain oil of cinnamon from cinnamon bark by steam distillation and observe the properties of cinnamaldehyde, the main chemical ingredient in oil of cinnamon. Discover the prized essence of a valuable spice!

Concepts

- Covalent bonding
- Sigma and pi bonds
- Single, double, and triple bonds
- Isomerism

- Soaps and soap-making
- Triglycerides
- Saponification
- Surfactants

- Ester functional group
- Carboxylic acids and alcohols
- Esterification reaction
- Equilibrium

- History of aspirin
- Salicylic acid derivatives
- Esters and esterification
- Excess and limiting reagents

- Essential oil
- Steam distillation
- Solvent extraction
- Aldehyde functional group

Experiment Summaries and Concepts

Demonstration

Cleaning with Charcoal—Turning Grape Juice into Water!

Miracle action! Wonder cleaning! No, it's not a TV commercial for a new cleaning product. It's activated charcoal, which has been used for thousands of years to remove contaminants from water and make it safe to drink. Demonstrate the uses of activated charcoal in both water treatment plants and products for the home by turning grape juice into water.

The Carbon Soufflé—Removing Water from Sugar!

Sugar is the main ingredient in this "chemical" soufflé. Sugars are carbohydrates—carbon plus water. Sulfuric acid, a powerful dehydrating agent, will dramatically remove the water from sugar, leaving only carbon behind. The exothermic reaction produces a great column of carbon that bubbles and grows out of the reaction beaker. This is a terrific demonstration to show the amount of chemical energy stored in food.

Feeling Blue—Organic Redox Reaction

Banish the blues from your chemistry classroom! The reaction of a reducing sugar with an organic redox indicator shows us how the blues may come and go when things get a little shaken up. This classic demonstration will help students develop good observation and reasoning skills. What causes the blue color to disappear and then reappear again? How long does it take for the blue color to disappear? What conditions will affect how fast the blue color dissipates?

Kaleidoscope . . . Optical Activity—Rotation of Plane Polarized Light

The existence of chiral compounds is a unique feature of natural products and the chemistry of life. Discover the interesting properties of "optically active" compounds using only polarizing films and a beaker of corn syrup on an overhead projector. When plane polarized light passes through a solution containing only one "handed" isomer of a chiral compound, the solution "rotates" the plane of polarized light. Optical rotation produces a kaleidoscope when many different colors of plane polarized light are used.

Salt-Out the Red, the White, and Blue—Making a Three-Layer Liquid

Create a beautiful three-layered liquid to demonstrate the salting-out effect and the relative density and miscibility of different solvents. The three liquid layers consist of toluene, methyl alcohol, and water. Methyl alcohol is "salted out" and separated from water by adding potassium carbonate. The water is dyed blue with copper(II) sulfate, and toluene is colored red with Sudan III. Why don't the methyl alcohol and toluene mix?

Concepts

- Activated charcoal
- Adsorption
- Water treatment

- Carbohydrates
- Dehydration reaction
- Exothermic reaction

- Oxidation–reduction
- Reducing sugar
- Redox indicator

- Chiral compounds
- Enantiomers
- Optical activity
- Plane polarized light

- Miscible and immiscible liquids
- Salting-out effect
- Density

Teacher Notes

Introduction to Functional Groups
Classifying Organic Compounds

Organic chemistry is the study of carbon and its compounds—their structures, properties, and chemical reactions. A tremendous number of compounds come under this heading. There are more than nine million organic compounds, more than seven million organic reactions associated with them! All organic compounds contain carbon, as well as hydrogen atoms attached to the carbon "skeleton" in predictable numbers. In most chemical reactions, the C—C skeleton does not change. Typically, organic reactions involve either C=C double or C≡C triple bonds in a molecule, or carbon atoms attached to "heteroatoms" such as oxygen, nitrogen or chlorine. Organic compounds are classified into functional group classes based on their structure and properties. A *functional group* is defined as a specific arrangement of atoms, such as —OH or —NH$_2$, that is responsible for the types of reactions an organic compound will undergo. Functional groups are the reactive groups in a molecule. They undergo characteristic chemical reactions, such as oxidation or dehydration, and also give related compounds similar physical properties. The study of organic functional groups provides the underlying basis for the science of organic chemistry. Knowing the chemistry of different functional groups allows chemists to explain and predict organic reactions and also to create new compounds.

Table 1 shows the structures of common organic functional groups. The symbol R is used to represent rings or chains of carbon atoms attached to the functional group.

Table 1. Structures of Organic Compounds and Functional Groups

Functional Group	Structure	Functional Group	Structure
Alkenes	C=C	Amines	R—NH$_2$ (or R$_2$NH and R$_3$N)
Alkynes	—C≡C—	Ketones	R—C(=O)—R'
Aromatic Compounds	(benzene ring)	Aldehydes	R—C(=O)—H
Alcohols	R—O—H	Carboxylic Acids	R—C(=O)—OH
Ethers	R—O—R'	Esters	R—C(=O)—O—R'
Alkyl Halides	R—X where X = F, Cl, Br, or I	Amides	R—C(=O)—NH$_2$

Introduction to Functional Groups

Teacher Notes

Name: ______________________________

Class/Lab Period: ______________________

Introduction to Functional Groups

The following examples illustrate the great variety of functional groups present in *natural products* (organic compounds isolated from nature).

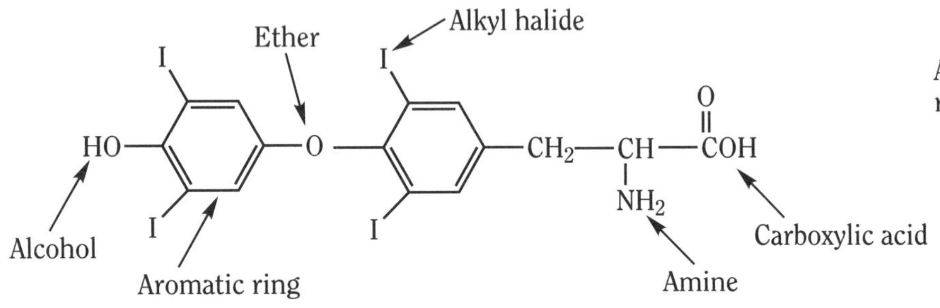

Thyroxine (Thyroid hormone)

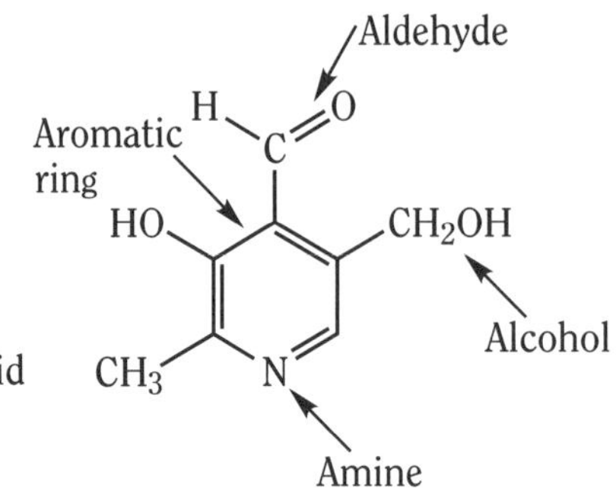

Pyridoxal (Vitamin B_6)

Circle and label the organic functional groups in the following natural and consumer products.

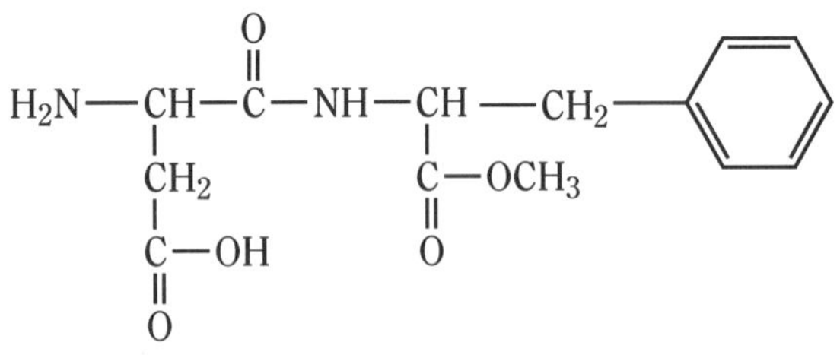

Aspartame (NutraSweet®)

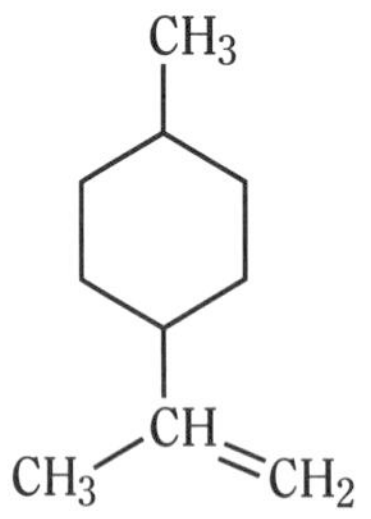

Limonene (Citrus peel oil)

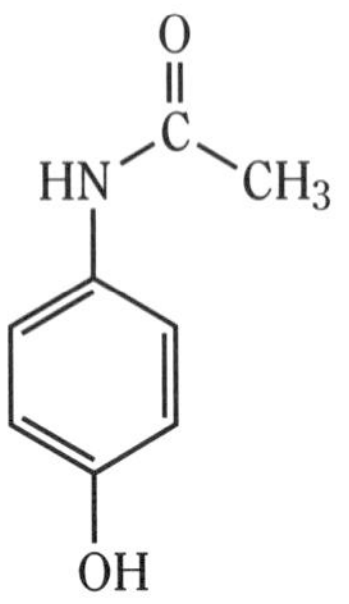

Acetaminophen (Tylenol®)

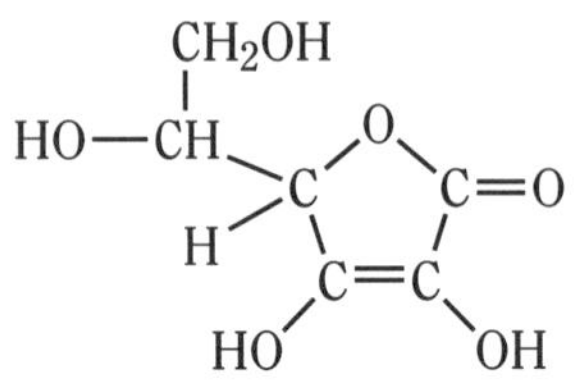

Ascorbic Acid (Vitamin C)

Teacher Notes

Naming Organic Compounds
The IUPAC System

"To call forth a concept, a word is needed." This quote from Antoine Lavoisier, known as the father of modern chemistry, describes the importance of learning the *language* of chemistry. The need for a comprehensive language of chemistry is especially evident in naming organic compounds. By the late 19th century, about 15,000 organic compounds were known. Today, the number is greater than nine million! Most organic compounds were originally known by common or "trivial" names that reflected the origin or source of the compound, but gave no information about its composition or structure. In 1892, an international group of chemists met in Geneva, Switzerland, and proposed a "systematic" method for naming organic compounds based on their structure. This method of naming organic compounds was adopted and further developed by the International Union of Pure and Applied Chemistry (IUPAC), an organization founded in 1919 to promote international cooperation in chemistry.

The goal of the IUPAC naming system is simple—one compound, one name. Every distinct compound should have a unique name, and there should be only one possible structure for any IUPAC name. The IUPAC method for naming organic compounds is a logical set of rules based on: (1) The longest "major" chain or ring of carbon atoms in a compound; (2) the presence of functional groups; and (3) the names of substituent groups or atoms attached to the main chain. The IUPAC name for a compound does not specify the number of hydrogen atoms. The number of hydrogen atoms in a compound and their locations are *assumed* based on the premise that every carbon atom has four covalent bonds. There are three parts to an IUPAC name:

The *root name* for an organic compound gives the number of carbon atoms in the "longest continuous carbon chain" that contains the functional group. The *prefix* identifies the number and location of atoms or groups attached to the longest chain, and the *ending* denotes the functional group. Table 1 gives the names of straight-chain alkanes C_nH_{2n+2} containing 1–10 carbon atoms. The root names are shown in boldface.

Table 1. Names of Straight-Chain Alkanes, C_nH_{2n+2}

n	Name	n	Name
1	**Meth**ane	6	**Hex**ane
2	**Eth**ane	7	**Hept**ane
3	**Pro**pane	8	**Oct**ane
4	**But**ane	9	**Non**ane
5	**Pent**ane	10	**Dec**ane

Name: _____

Class/Lab Period: _____

Naming Organic Compounds Worksheet

Table 2. IUPAC Names of C₅H₁₂ Isomers

Structure	CH₃—CH₂—CH₂—CH₂—CH₃	CH₃—CH—CH₂—CH₃ with CH₃ on C-2 (numbered 1 2 3 4)	CH₃—C(CH₃)(CH₃)—CH₃ (numbered 1 2 3)
Root Name	Pent-	But-	Prop-
Prefix(es)	None	Methyl- (located on C–2)	Dimethyl- (both located on C–2)
Ending	-ane	-ane	-ane
IUPAC Name	Pentane	2-Methylbutane	2,2-Dimethylpropane

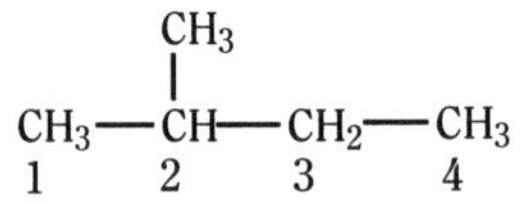

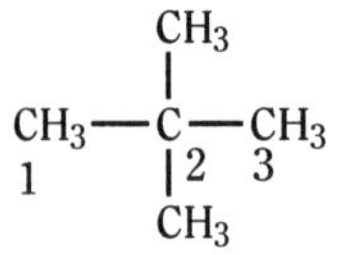

Table 2 shows the steps involved in naming isomers of "pentane" (C_5H_{12}). IUPAC names are generally one word. Carbon groups attached to the "main" chain are called alkyl groups and their names are derived from the corresponding alkane. Thus, a methyl group is CH_3—, an ethyl group is CH_3CH_2—, a propyl group is $CH_3CH_2CH_2$—, etc. An isopropyl group is a branched group, $(CH_3)_2CH$—. The location of an alkyl group or other atom is specified with a number, which is separated from the name by a hyphen. The prefixes for halogen atoms are fluoro-, chloro-, bromo-, etc. The ending in an IUPAC name specifies the functional group. Some examples of functional group endings are –ene for alkenes containing a C═C double bond, –yne for alkynes containing a C≡C triple bond, and –ol for alcohols containing an —OH group.

The following compounds provide additional examples of the IUPAC naming system:

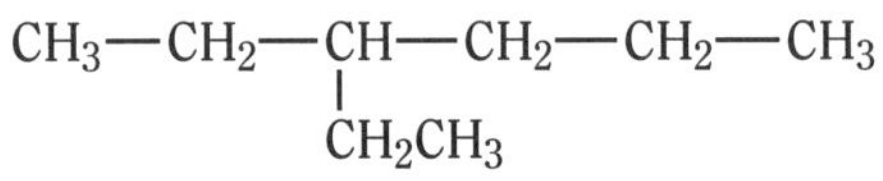

3-Ethylhexane

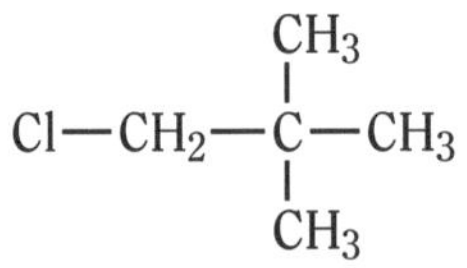

1-Chloro-2,2-dimethylpropane

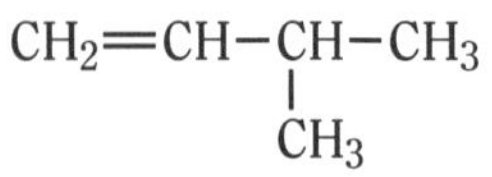

3-Methyl-1-butene*

Name the following compounds using the IUPAC system.

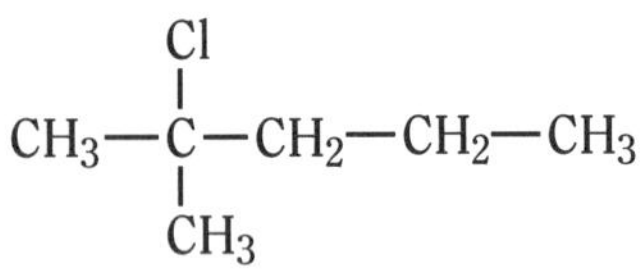

A

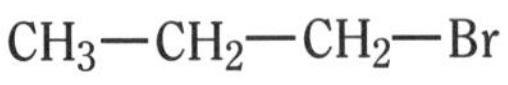

B

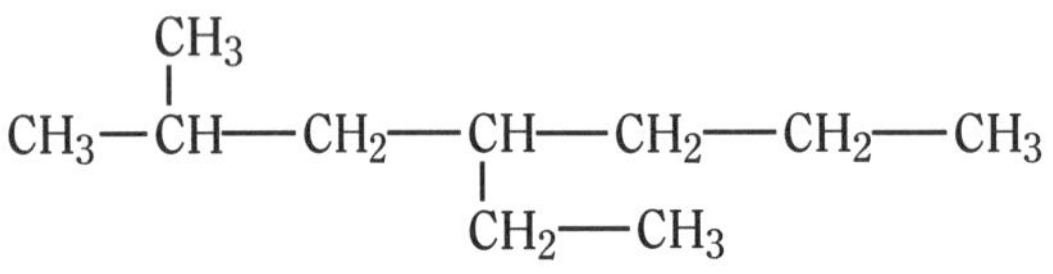

C

*1– Denotes the position of the double bond.

Teacher Notes

Answers to naming exercises:

A
2-chloro-2-methylpentane

B
1-bromopropane

C
2-methyl-4-ethylheptane

Models of Organic Compounds
Guided Inquiry

Teacher Notes

Introduction

There are at least nine million organic compounds. What factors are responsible for the tremendous number of organic compounds? What makes all of these compounds different? Building organic molecules using models can help us understand the basic structures of organic compounds.

Concepts

- Covalent bonding
- Sigma and pi bonding
- Single, double, and triple bonds
- Isomerism

Background

The term organic chemistry refers to the study of compounds containing carbon. The name reflects the historical roots of organic chemistry—it was thought that compounds obtained from living organisms required a "vital force" for their existence. This notion was discarded in 1828, when the first organic compound was synthesized in the lab, but the name remains.

Carbon is unique among the elements because of the large number and diverse structures of compounds that it forms. Several factors help explain why compounds containing carbon are well suited to the chemistry of life:

- Carbon forms strong and stable bonds with other carbon atoms. The ability of carbon to form strong C–C bonds of almost infinite chain length is called *catenation*.

- Chains of carbon atoms can "close in" on themselves to form rings in addition to chains. Many different ring sizes are possible, but five-, six-, and seven-membered rings are the most common.

- The electronegativity of carbon (2.5) is in the middle of the range of values for all elements (0.7 – 4.0). This means that carbon forms strong covalent bonds with nonmetals and even many metals, from aluminum to zirconium.

- The valency of carbon is four—carbon forms four covalent bonds to achieve a closed shell electron configuration (a stable octet). This is the maximum number of bonds a second row element can form.

- Because of their small size, carbon atoms form strong multiple bonds (double and triple bonds) to other carbon atoms, as well as to nitrogen, oxygen, and sulfur atoms. The strength of pi bonds in double and triple bonds depends on the sizes of the atoms.

Activity Overview

The purpose of this activity is to build organic molecules using models. The models will be used to draw structural formulas of organic compounds, determine the general formulas for different classes of hydrocarbons, and develop the concept of isomers of organic compounds.

A quick review of drawing Lewis structures is in order before beginning this activity! Recall also the difference between sigma and pi bonds and the VSEPR theory for predicting the shapes of molecules.

Models of Organic Compounds – *Page 2*

Name: _____

Class/Lab Period: _____

Models of Organic Compounds Worksheet

Materials

Organic model set with at least six carbon atoms

Review Questions

1. What is the maximum number of covalent bonds each of the following elements will form in a neutral compound?

   Hydrogen _____   Carbon _____   Nitrogen _____

   Oxygen _____   Fluorine _____   Boron _____

2. The *structural formula* of a molecule shows all of the atoms in the structure and the order in which they are connected by covalent bonds. Add hydrogen atoms as needed to each atom in the following structural formulas so that each atom has a closed shell electron configuration and zero charge.

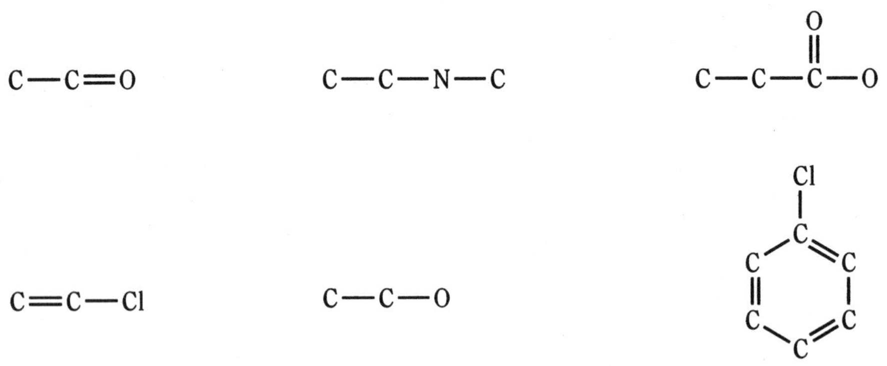

   C—C=O   C—C—N—C   C—C—C(=O)—O

   C=C—Cl   C—C—O   [benzene ring with Cl substituent]

3. Methane, CH_4, is the chief component of natural gas. (a) What is the molecular geometry around the carbon atom in methane? (b) What is the H–C–H bond angle in methane? (c) Draw a diagram that illustrates the three-dimensional shape of methane.

Structure of Organic Compounds

1. Build models of ethane, C_2H_6, and propane, C_3H_8, and write out their structural formulas.

Page 3 – **Models of Organic Compounds**

Teacher Notes

2. Do the C–C single bonds in ethane and propane rotate freely? Explain.

3. There are two possible structures for butane, C_4H_{10}. Build models of both structures and draw their structural formulas.

4. The two possible structural formulas for butane are called *isomers*. Write a general definition of isomers that describes the relationship between the two structures.

5. Without building models, draw out the possible structural formulas for three isomers of pentane, C_5H_{12}.

6. Alkanes are hydrocarbons—compounds containing only carbon and hydrogen—in which all of the C–C bonds are single bonds. What is the *general formula* for an alkane, where *n* is the number of carbon atoms?

The common names for the straight-chain and branched-chain "butane" isomers in Question #3 are butane and isobutane, respectively. The "pentane" isomers in Question #4 are also known by common names, pentane, isopentane, and neopentane. See the "Naming Organic Compounds" worksheet for IUPAC rules.

7. *Alkenes* are hydrocarbons that contain at least one C=C double bond in their structure. Build models of ethene (C_2H_4) and propene (C_3H_6) and draw their structural formulas.

8. Describe the molecular geometry around the C=C double bond in an alkene. What is the H–C–H bond angle in ethene?

Models of Organic Compounds

Models of Organic Compounds – Page 4

9. Unlike C–C single bonds, C=C double bonds do not rotate. Draw a diagram showing the overlap of the orbitals responsible for the sigma and pi bonds, respectively, in a C=C double bond. Use the orbital diagram to explain why the C=C double bond does not freely rotate.

10. Butene (C_4H_8) has one C=C double bond in the structure. Draw structures for three possible *structural isomers* of butene.

11. The structural formula for *2-butene* can be abbreviated CH_3–CH=CH–CH_3. Because of the lack of free rotation about the C=C double bond (see Question 9), there are two possible structures for this compound. Build models and draw structural formulas for two *three-dimensional structures* of 2-butene.

12. The two isomers of 2-butene shown in Question 11 are called *geometric isomers*. What is the same and what is different about geometric isomers?

13. What is the general formula of an alkene, where *n* is the number of carbon atoms? Why do you think alkenes are called unsaturated and alkanes are called saturated hydrocarbons?

14. Benzene, C_6H_6, is the "parent" compound of a class of compounds called aromatic compounds that are very common in nature. The carbon "skeleton" for benzene is shown below. Add *hydrogen atoms* and *double bonds*, as necessary, to complete the structure of benzene.

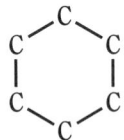

Teacher Notes

Page 5 – **Models of Organic Compounds**

15. Build a model of benzene and describe its general structure (planar, tetrahedral, etc.)

16. The structural formula of benzene shown in Question 14 has alternating single (C–C) and double (C=C) bonds. It has been found, however, that all of the carbon-carbon bonds in benzene are identical! This fact may be explained in terms of resonance. Define *resonance* and draw two resonance forms for benzene.

17. Alcohols are organic compounds containing an –OH group attached to a carbon atom. Draw the structural formula of ethyl alcohol, C_2H_5OH.

18. Low molecular weight alcohols such as ethyl alcohol are polar compounds and are miscible with water. As the number of carbon atoms in an alcohol increases, the solubility of the alcohol in water decreases. Thus, octanol, $C_8H_{17}OH$, is practically insoluble in water. Explain.

19. Compounds containing at least one carbon atom that is attached to four different groups give rise to a special class of isomers called enantiomers. *Enantiomers* are defined as non-superimposable mirror images of each other. Build models and complete the following diagrams to show the enantiomers of *alanine,* an amino acid.

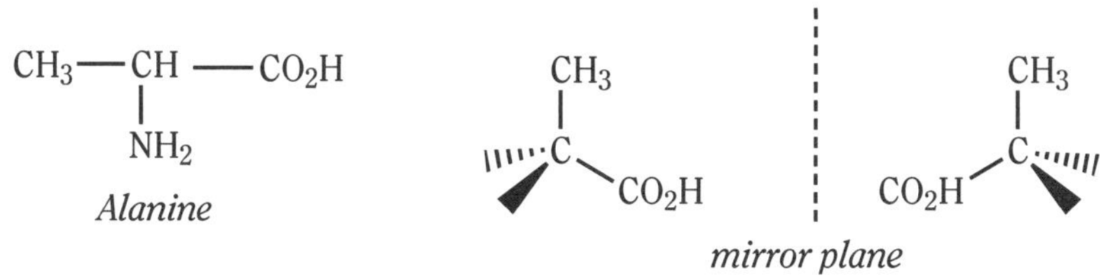

CH_3—CH—CO_2H
　　　|
　　　NH_2
Alanine

mirror plane

20. What does it mean to say that the enantiomers shown in Question 19 are *non-superimposable?* Why do you think this property of molecules is sometimes called "handedness?"

Models of Organic Compounds

Teacher's Notes
Models of Organic Compounds

Master Materials List *(for a class of 30 students working in groups of three)*

Organic model sets with at least six carbon atoms, 10*

*See the "Organic Small Group Model Set" available from Flinn Scientific Inc., Catalog No. AP5453.

Hints for Guided-Inquiry Activities

- In general, guided-inquiry activities are most successful if students understand that the activity replaces the lecture. Students are more likely to take responsibility for learning when they are actively engaged in the process of "constructing knowledge." Guided-inquiry activities simulate the scientific method—students look at data, search for patterns or relationships, and try to identify guiding principles that will explain the data.

- The teacher's role in guided-inquiry activities is very important. The atmosphere in the classroom must be conducive to independent learning. The teacher facilitates learning by monitoring students, keeping them on track, and reviewing progress at key junctures. In this activity, for example, the teacher may want to call a time-out after Question 4, the definition of isomers. Ask several groups to give their definitions, then ask students to explain or defend their definition, or to modify their definition based on new information.

- The following topics and concepts should be reviewed prior to scheduling this activity: Lewis structures, VSEPR theory, multiple (double and triple) bonds, and hybrid orbitals.

- Protocols for teaching Lewis structures that work well in general chemistry (count up all the valence electrons, add single bonds, followed by lone pairs, etc.) are not as useful when teaching organic chemistry. An alternative strategy is to teach students the typical number of bonds that an atom will form when it has zero formal charge. Carbon forms four bonds, nitrogen three bonds, oxygen two bonds, fluorine one bond.

- Students need good spatial reasoning skills to visualize molecules in three dimensions. All students will benefit from the opportunity to hold models in their hands, rotate them, turn them upside down, etc.

Teacher Notes

Models of Organic Compounds Worksheet
Answer Key

Student answers will vary.

Review Questions

1. What is the maximum number of covalent bonds each of the following elements will form in a neutral compound?

   Hydrogen __1__   Carbon __4__   Nitrogen __3__

   Oxygen __2__   Fluorine __1__   Boron __3__

2. The *structural formula* of a molecule shows all of the atoms in the structure and the order in which they are connected by covalent bonds. Add hydrogen atoms as needed to each atom in the following structural formulas so that each atom has a closed shell electron configuration and zero charge.

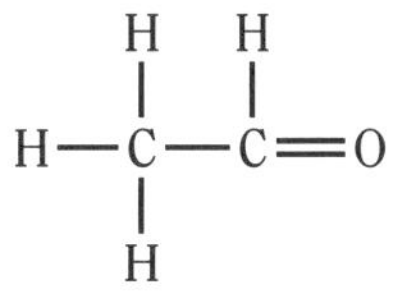

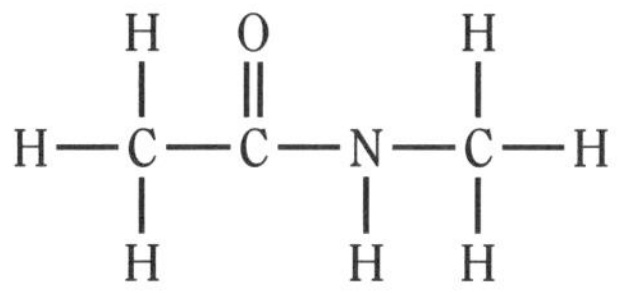

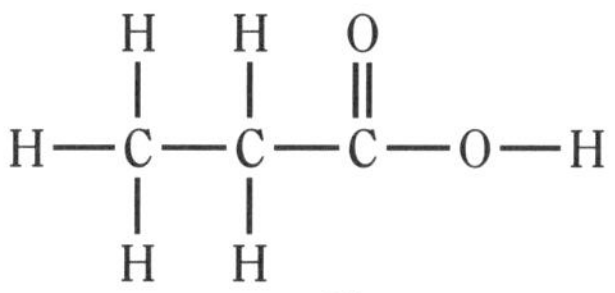

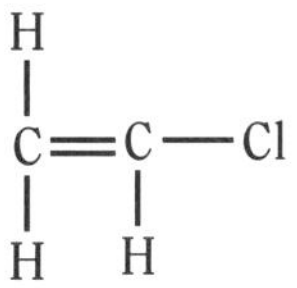

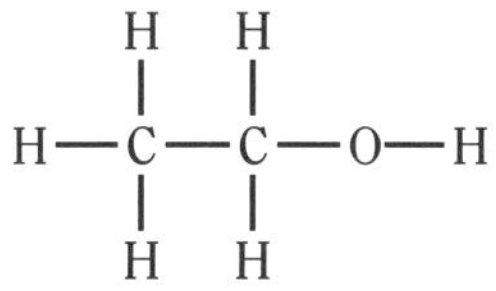

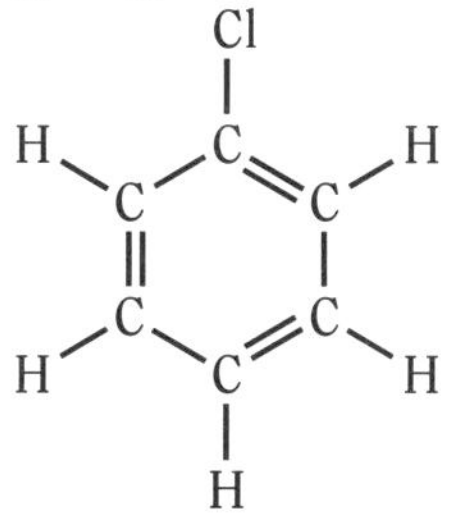

3. Methane, CH_4, is the chief component of natural gas. (a) What is the molecular geometry around the carbon atom in methane? (b) What is the H–C–H bond angle in methane? (c) Draw a diagram that illustrates the three-dimensional shape of methane.

   (a) Methane has a tetrahedral geometry around the central carbon atom.

   (b) The ideal tetrahedral H–C–H bond angle is 109.5°.

   (c)

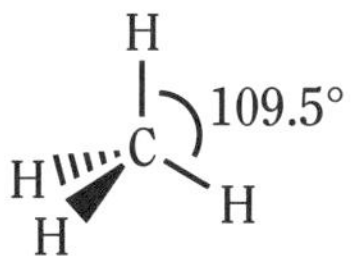

Structure of Organic Compounds

1. Build models of ethane, C_2H_6, and propane, C_3H_8, and write out their structural formulas.

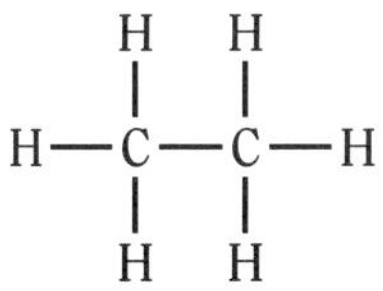

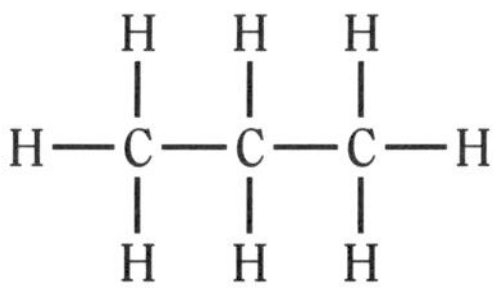

Models of Organic Compounds

Teacher's Notes

2. Do the C–C single bonds in ethane and propane rotate freely? Explain.

   Yes, the hydrogen atoms on adjacent carbon atoms in ethane can "slide" past each other as the C–C bond turns or rotates.

3. There are two possible structures for butane, C_4H_{10}. Build models of both structures and draw their structural formulas.

4. The two possible structural formulas for butane are called *isomers*. Write a general definition of isomers that describes the relationship between the two structures.

   Isomers have the same molecular formulas but different structural formulas.

5. Without building models, draw out the possible structural formulas for three isomers of pentane, C_5H_{12}.

6. Alkanes are hydrocarbons—compounds containing only carbon and hydrogen—in which all of the C–C bonds are single bonds. What is the *general formula* for an alkane, where n is the number of carbon atoms?

   C_nH_{2n+2}

7. *Alkenes* are hydrocarbons that contain at least one C=C double bond in their structure. Build models of ethene (C_2H_4) and propene (C_3H_6) and draw their structural formulas.

8. Describe the molecular geometry around the C=C double bond in an alkene. What is the H–C–H bond angle in ethene?

   The molecular geometry about the C=C bond is planar—the two carbon atoms and atoms attached to them lie in a plane. The H–C–H bond angle is 120°.

Teacher Notes

Teacher's Notes

9. Unlike C–C single bonds, C=C double bonds do not rotate. Draw a diagram showing the overlap of the orbitals responsible for the sigma and pi bonds, respectively, in a C=C double bond. Use the orbital diagram to explain why the C=C double bond does not freely rotate.

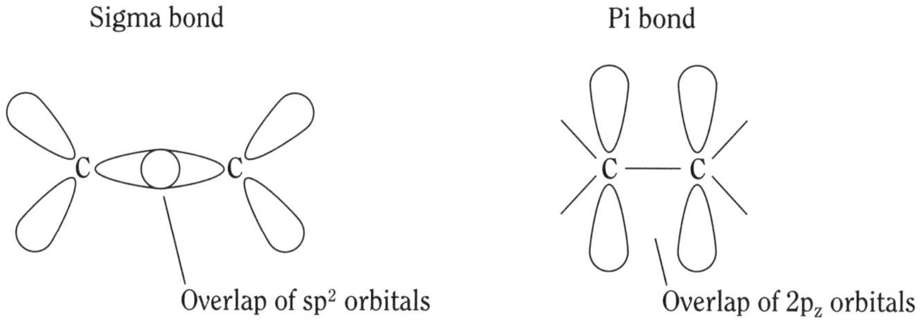

   Sigma bond   Pi bond

   Overlap of sp² orbitals   Overlap of 2p_z orbitals

   "Turning" the C=C bond would destroy the overlap of the p orbitals in the pi bond.

10. Butene (C_4H_8) has one C=C double bond in its structure. Draw structures for three possible *structural isomers* of butene.

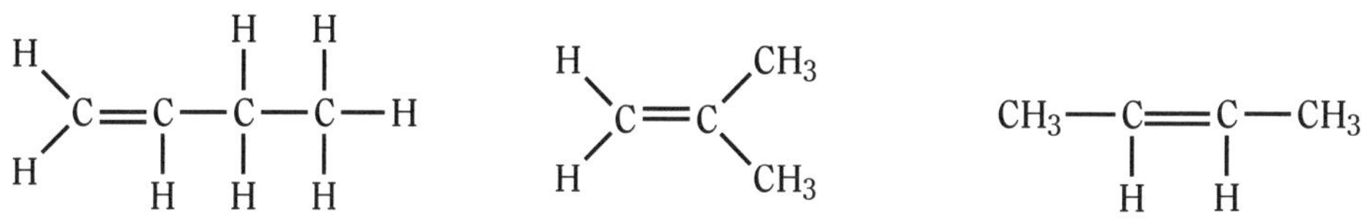

11. The structural formula for *2-butene* can be abbreviated CH_3–CH=CH–CH_3. Because of the lack of free rotation about the C=C double bond (see Question 9), there are two possible structures for this compound. Build models and draw structural formulas for two *three-dimensional structures* of 2-butene.

12. The two isomers of 2-butene shown in Question 11 are called *geometric isomers*. What is the same and what is different about geometric isomers?

    Geometric isomers have the same molecular formula and the same structural formula, but different arrangements of atoms in space.

13. What is the general formula of an alkene, where *n* is the number of carbon atoms? Why do you think alkenes are called unsaturated and alkanes are called saturated hydrocarbons?

    C_nH_{2n}

    Alkenes are "unsaturated" because they contain fewer than the maximum number of hydrogen atoms possible for the number of carbon atoms. Alkanes are "saturated" because they cannot add any more hydrogen atoms to their structures.

Models of Organic Compounds

Teacher's Notes

14. Benzene, C_6H_6, is the "parent" compound of a class of compounds called aromatic compounds that are very common in nature. The carbon "skeleton" for benzene is shown below. Add *hydrogen atoms* and *double bonds,* as necessary, to complete the structure of benzene.

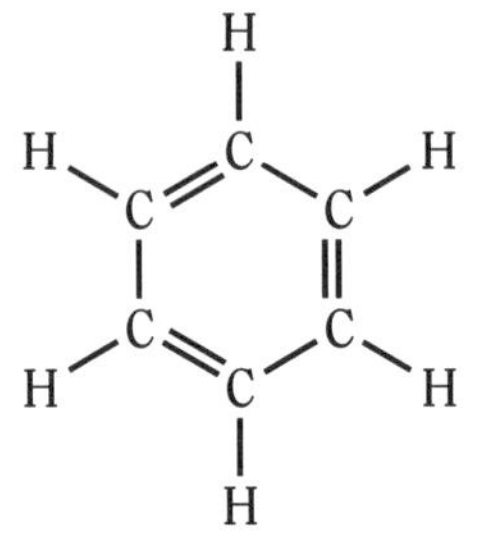

15. Build a model of benzene and describe its general structure (planar, tetrahedral, etc.).

    Benzene is a planar molecule—all of the atoms lie in a single plane.

16. The structural formula of benzene shown in Question 14 has alternating single (C–C) and double (C=C) bonds. It has been found, however, that all of the carbon-carbon bonds in benzene are identical! This fact may be explained in terms of resonance. Define *resonance* and draw two resonance forms for benzene.

    Resonance occurs in a molecule when it is possible to write two or more valid Lewis structures for the molecule. The actual structure of benzene is the average of the two possible Lewis structures.

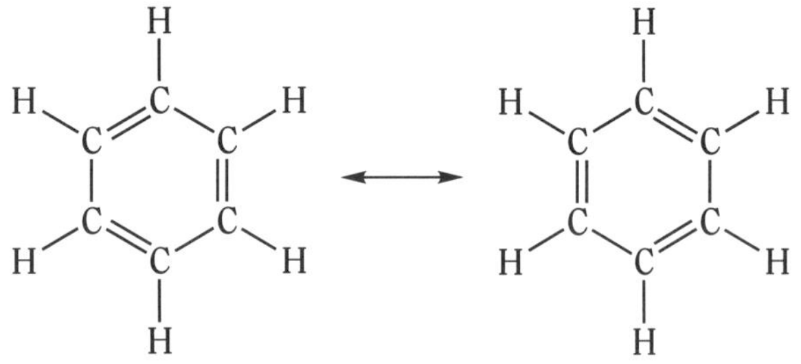

17. Alcohols are organic compounds containing an –OH group attached to a carbon atom. Draw the structural formula of ethyl alcohol, C_2H_5OH.

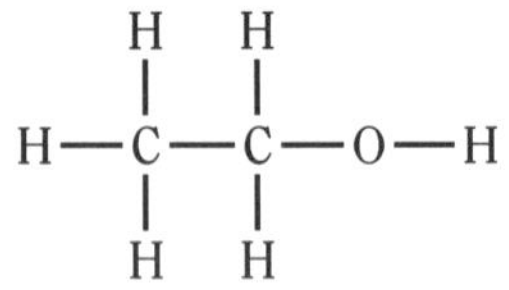

18. Low molecular weight alcohols such as ethyl alcohol are polar compounds and are miscible with water. As the number of carbon atoms in an alcohol increases, the solubility of the alcohol in water decreases. Thus, octanol, $C_8H_{17}OH$, is practically insoluble in water. Explain.

    In octanol, the polar –OH group "competes" with the nonpolar chain of eight carbon atoms. It is insoluble in water because the long nonpolar chain predominates.

Flinn ChemTopic™ Labs — Chemistry of Organic Compounds

Teacher's Notes

Teacher Notes

19. Compounds containing at least one carbon atom that is attached to four different groups give rise to a special class of isomers called enantiomers. *Enantiomers* are defined as non-superimposable mirror images of each other. Build models and complete the following diagrams to show the enantiomers of *alanine,* an amino acid.

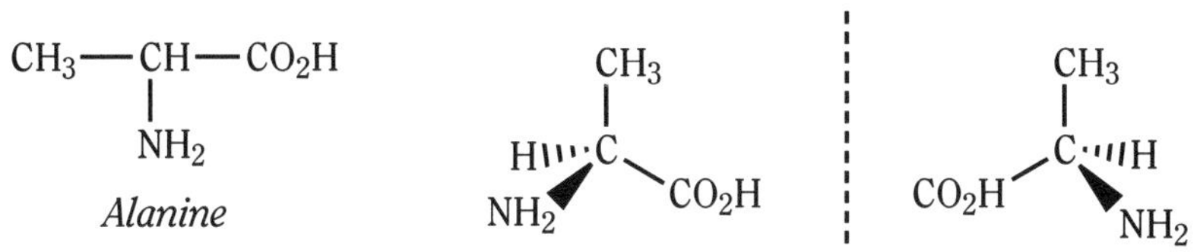

CH_3—CH—CO_2H
    |
    NH_2
Alanine

[Structure showing: CH_3 up, H wedge, NH_2, CO_2H — C center] | [Mirror image structure: CH_3 up, CO_2H, H wedge, NH_2 — C center]

mirror plane

20. What does it mean to say that the enantiomers shown in Question 19 are *non-superimposable?* Why do you think this property of molecules is sometimes called "handedness?"

The enantiomers cannot be superimposed on one another by rotating or turning the molecules. Our hands are an example of non-superimposable mirror images.

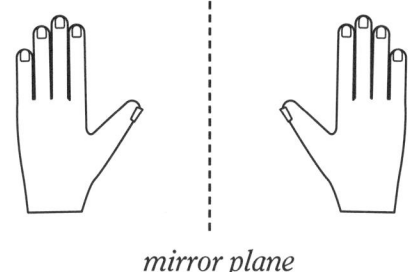

mirror plane

15   Models of Organic Compounds

Teacher's Notes

Making Soap
The Oldest Organic Reaction

Teacher Notes

Introduction

Soap-making is an ancient craft and one of the oldest known chemical reactions involving organic compounds. Soaps are sodium and potassium salts of fatty acids. They are prepared by reacting fats and oils with a strong base, such as sodium hydroxide or potassium hydroxide.

Concepts

- Soaps and soap-making
- Triglycerides
- Saponification
- Surfactants

Background

Legend has it that the word *soap* comes from Mount Sapo, a hill in Rome that was the site of animal sacrifice. According to the legend, animal fat and ashes washed down the mountain with the rain, producing sudsy river water that was used to wash clothes. It's an interesting story, but there is no factual evidence for this legend. The earliest written reference comes from the Roman historian Pliny the Elder in the first century A.D. Pliny described the preparation of *sapo* from goat fat and wood ashes and attributed the invention to the Gauls, who used it for hair treatment rather than for bathing or cleaning. Historical references to soap may be found in ancient Babylonian and Egyptian artifacts dating as far back as 2500 B.C.

Soap-making is also associated with colonial America and pioneers on the American frontier. The soap was made by boiling fat with a concentrated solution of potash (potassium carbonate) extracted from wood ashes with hot water. Potassium carbonate solutions are *caustic*—strongly basic and irritating to the skin and eyes. Soap made in this way was likely to contain excess (unreacted) potassium carbonate and was therefore quite harsh, leaving the skin rough and dry. This frontier method of soap-making may appear primitive, but it is still used in almost the same form today to make both commercial and handmade soaps. The methods are safer, however, and soaps are milder because the starting materials are pure, the chemistry is well-understood, and the reactants can be mixed in the right ratio.

The process of making soap is called *saponification* and is one of the earliest examples of using organic chemistry to produce a man-made product. Saponification involves the reaction of *triglycerides*—natural fats and oils—with sodium or potassium hydroxide.

$$\begin{array}{l} CH_2-O-\overset{\overset{O}{\|}}{C}-(CH_2)_{16}CH_3 \\ | \\ CH-O-\overset{\overset{O}{\|}}{C}-(CH_2)_{12}CH_3 \\ | \\ CH_2-O-\overset{\overset{O}{\|}}{C}-(CH_2)_7CH=CH(CH_2)_7CH_3 \end{array}$$

Figure 1. Structure of a Triglyceride.

Fats and oils are all triglycerides. The distinction between a fat and an oil rests on the source of the triglyceride—fats are obtained from animal sources, oils from plants. Most, but not all, oils are liquids at room temperature because they contain a greater proportion of unsaturated fatty acids. (Introducing double bonds into a triglyceride puts "bends" or "kinks" in the shape of the molecule, which lowers the melting point.) The exceptions to this general rule are coconut oil and palm oil, which contain relatively low molecular weight saturated fatty acids.

Triglycerides are *esters* containing three fatty acid groups attached via ester linkages to a glycerol backbone (Figure 1). The products of a saponification reaction are sodium or potassium salts of fatty acids and glycerol (Equation 1).

$$\begin{array}{c}\text{CH}_2-\text{O}-\overset{\text{O}}{\underset{\|}{\text{C}}}-\text{R} \\ | \\ \text{HC}-\text{O}-\overset{\text{O}}{\underset{\|}{\text{C}}}-\text{R}' \\ | \\ \text{CH}_2-\text{O}-\overset{\text{O}}{\underset{\|}{\text{C}}}-\text{R}''\end{array} + 3\text{NaOH} \longrightarrow \begin{array}{c}\text{H} \\ | \\ \text{H}-\text{C}-\text{OH} \\ | \\ \text{H}-\text{C}-\text{OH} \\ | \\ \text{H}-\text{C}-\text{OH} \\ | \\ \text{H}\end{array} + \begin{array}{c}\text{Na}^+\ ^-\text{O}-\overset{\text{O}}{\underset{\|}{\text{C}}}-\text{R} \\ \text{Na}^+\ ^-\text{O}-\overset{\text{O}}{\underset{\|}{\text{C}}}-\text{R}' \\ \text{Na}^+\ ^-\text{O}-\overset{\text{O}}{\underset{\|}{\text{C}}}-\text{R}''\end{array} \qquad Equation\ 1$$

Triglyceride   *Sodium Hydroxide*   *Glycerol*   *Sodium salts of long-chain fatty acids (soap)*

Most fats and oils contain a mixture of fatty acid residues of different chain lengths. The most common fatty acids have 12–18 carbon atoms and may be saturated or unsaturated. Unsaturated and polyunsaturated fatty acids contain one or more C=C double bonds, respectively, in their structures, while saturated fatty acids contain no C=C double bonds.

Soaps belong to a class of compounds called *surface-active agents* or *surfactants,* which also include detergents and emulsifying agents. A *surfactant* is defined as a compound that reduces surface tension when dissolved in water or in aqueous solutions. All surfactants have two basic features in common. One end of a surfactant molecule is usually a long, nonpolar hydrocarbon chain, resembling a "tail." The hydrocarbon tail is said to be *hydrophobic* (water-fearing) because it tends to repel or exclude water and will not dissolve in water. The other end of a surfactant molecule is a small ionic or polar group that is hydrophilic (water-loving). The hydrophilic group will tend to be solvated or surrounded by water molecules and will dissolve in water. These two competing structural features give soaps and other surfactants their unique properties.

When dissolved in water, soaps and other surfactant molecules spontaneously self-associate to form spherical aggregates called *micelles* (Figure 2).

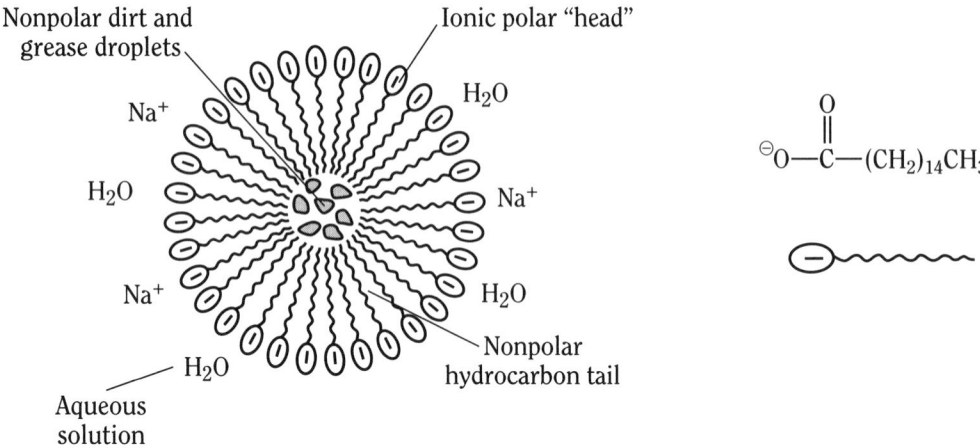

Figure 2. Structure and Properties of a Micelle.

Micelles are spherical, three-dimensional aggregates.

Page 3 – **Making Soap**

Teacher Notes

The nonpolar hydrocarbon tails in the soap molecules spontaneously arrange themselves toward the interior of the micelle, giving it a hydrophobic core that repels and thus excludes water. The ionic head groups are arranged on the outside surface of the micelle and are surrounded by water molecules. The ability of soap molecules to form micelles explains how and why soaps work. Dirt and grease are nonpolar, hydrophobic substances that are not soluble in water. If water alone were used for washing or cleaning, the hydrophobic dirt and grease molecules would not dissolve in the water. In soapy water, however, dirt and grease molecules become trapped or suspended within the hydrophobic core of a micelle. The soap thus disperses or breaks up the dirt particles and dissolves them in the water. The dirt-containing micelles are water-soluble and are rinsed away in the wash. The formation of micelles is also related to the *emulsifying action* of soaps—their ability to form stable mixtures or suspensions of two or more immiscible liquids.

Experiment Overview

The purpose of this experiment is to make soap and study its properties. The soap will be prepared via saponification of a fat and oil with a solution of sodium hydroxide. The properties of the soap will then be investigated—its pH, texture, emulsifying action, and solubility in hard water.

Pre-Lab Questions

1. Define the following terms: Triglyceride, saponification, surfactant, micelle.

2. What is the principal safety hazard in this experiment? Explain why soap made by primitive methods was likely to be very harsh.

3. Olive oil has a *saponification value* of 190 mg KOH per gram. This means that it takes 190 mg of potassium hydroxide to react completely with 1 g of olive oil.

    (a) What is the mole ratio for the reaction of a triglyceride with KOH? (See Equation 1.)

    (b) Divide 0.190 g of KOH by its molar mass to calculate the number of moles of KOH that will react with 1 g of olive oil.

    (c) Use the answers to (a) and (b) to determine the number of moles corresponding to 1 g of olive oil. Divide 1 g by this number of moles to calculate the average molar mass (g/mole) of olive oil.

Materials

Calcium chloride solution, $CaCl_2$, 0.5 M, 1 mL
Iron(III) chloride solution, $FeCl_3$, 0.5 M, 1 mL
Magnesium chloride solution, $MgCl_2$, 0.5 M, 1 mL
Olive oil or vegetable oil, 3 g
Sodium hydroxide solution, NaOH, 6 M, 5.8 mL
Tallow or lard (solid fat), 7 g
Hot plate
Wash bottle and distilled water
Weighing dishes, small, 2
Balance, 0.1-g precision

Beakers, 50- and 250-mL
Beral-type pipets, 4
Graduated cylinder, 10-mL
pH meter, handheld, or pH paper
Spatula, metal
Stirring rod
Test tubes, small, 5
Test tube rack
Thermometer

19 Making Soap

Making Soap – Page 4

Safety Precautions

Sodium hydroxide solution is extremely corrosive. It can cause severe skin burns and is especially dangerous to the eyes. Notify your teacher of any spills and clean up all spills immediately. Avoid contact of all chemicals with eyes and skin. Do not use the soap product or remove it from the lab unless permission is granted by the teacher. Wear chemical splash goggles, chemical-resistant gloves, and a chemical-resistant apron. Wash hands thoroughly with soap and water before leaving the lab.

Procedure

Part A. Preparation of Soap

1. Tare a 50-mL beaker on the electronic balance. Place about 7 g of tallow (solid fat) into the beaker using a metal spatula. The fat is messy to work with—try not to get any on the outside of the beaker.

2. Measure about 3 g of olive oil into the same 50-mL beaker.

3. Place the beaker on a hot plate *at the lowest setting* or inside a 250-mL beaker filled with hot tap water. Heat the contents of the beaker until the fat melts and the olive oil and fat form a homogeneous solution.

4. Remove the reaction beaker from the hot plate or warm water bath.

5. Measure 5.8 mL of 6 M sodium hydroxide solution into a 10-mL graduated cylinder and *carefully* add the sodium hydroxide solution to the melted fat and oil mixture. Heat gently in hot water to about 40 °C.

6. Stir the reaction mixture with a stirring rod. After 5 minutes, place the reaction beaker into a larger, 250-mL beaker filled with cold water.

7. Continue stirring until the soap mixture gets thick—the product is ready to pour when the soap that drips back into the beaker from the stirring rod will trace a path (mark a trail) on the surface.

8. Label two small weighing dishes with your initials. Carefully pour the thickened soap solution from the reaction beaker into the weighing dishes.

9. Allow the soap to dry (cure) for several days. In the data table, describe the color, texture, and appearance of the soap.

Part B. Properties of Soap

10. Add about 40 mL of distilled water to the leftover soap on the sides and bottom of the beaker. Gently scrape the soap into the water as needed.

11. Heat the beaker on a hot plate at a medium setting and stir gently until the leftover soap dissolves. Cool to room temperature. In the data table, describe the appearance of the soap solution.

12. Measure the pH of the soap solution using either a handheld pH meter or pH paper. In the data table, record the pH of the solution.

Teacher Notes

The amount of sodium hydroxide used in step 5 should be precisely measured in order to avoid having excess base in the soap product. The soap is ready to pour in step 8 when it has the consistency of pea soup or honey. Waiting too long, however, before pouring the soap will result in a grainy-textured product. (The soap will not be smooth.)

Teacher Notes

13. Place three test tubes in a test tube rack and label them A–C. Pour about 3 mL of the soap solution (step 11) into each test tube.

14. Add 5 drops of calcium chloride solution to test tube A, 5 drops of iron(III) chloride solution to test tube B, and 5 drops of magnesium chloride solution to test tube C. Swirl each test tube to mix the contents. Describe the color and appearance of the mixture in each test tube in the data table.

15. Place two clean test tubes in the test tube rack. Add 3 mL of distilled water to the first test tube and 3 mL of the soap solution (step 11) to the second test tube.

16. Add one drop of olive oil to each test tube. Swirl or shake each test tube and then let sit for 5 minutes.

17. Describe the observations of this *emulsification* test in the data table.

18. When the soap bars are dry, measure and record the pH of the soap. Place a drop of water on pH paper and press the paper onto the top of the soap bar.

Compare the properties of the soap in Part B with a variety of commercial brands, including both liquid soap and bar soap.

Name: _____

Class/Lab Period: _____

Making Soap

Data Table

Color, Texture and Appearance of Soap	
pH of Soap (after one week)	
pH and Appearance of Soap Solution	
Hard Water Tests — Ca^{2+} Ions	
Hard Water Tests — Fe^{3+} Ions	
Hard Water Tests — Mg^{2+} Ions	
Emulsification Test	

Post-Lab Questions *(Use a separate sheet of paper to answer the following questions.)*

1. Compare the color, texture, and appearance of the homemade soap versus your favorite brand of hand soap.

2. Is the homemade soap solution acidic or basic? Explain.

3. How do the reactions of the soap solution with calcium, iron, and magnesium ions mimic what happens when soap is used in hard water? Draw the structure for the product obtained in the reaction of soap with calcium ions. Describe the most common result seen around the home due to the reaction of soap with hard water.

4. Compare the results of the emulsification test of olive oil with water and with soap solution. Explain in terms of the ability of soap to form micelles.

5. Explain why most soaps contain glycerol and describe the properties that glycerol adds to soap.

6. Draw the general structure of a triglyceride containing both saturated and unsaturated fatty acids. Circle and label the glycerol backbone, an ester functional group, and an unsaturated fatty acid residue.

7. Most homemade recipes for making soap use about 5% excess fat. Explain the benefits of using excess fat to make soap.

Teacher Notes

Teacher's Notes
Making Soap

Master Materials List *(for a class of 30 students working in pairs)*

Calcium chloride solution, $CaCl_2$, 0.5 M, 15 mL	Beakers, 50-mL, 15
Iron(III) chloride solution, $FeCl_3$, 0.5 M, 15 mL	Beakers, 250-mL, 15
Magnesium chloride solution, $MgCl_2$, 0.5 M, 15 mL	Beral-type pipets, 60
Olive oil or vegetable oil, 45 g	Graduated cylinders, 10-mL, 15
Sodium hydroxide solution, NaOH, 6 M, 100 mL	pH meter or pH paper
Tallow or lard (solid fat), 105 g	Spatulas, 15
Hot plates, 3–5*	Stirring rods, 15
Wash bottles and distilled water, 15	Test tubes, small, 75
Weighing dishes, small, 30	Test tube racks, 15
Balances, 0.1-g precision, 3*	Thermometers, 15

*Students may share balances and hot plates.

Preparation of Solutions *(for two classes of 30 students working in pairs)*

Calcium Chloride, 0.5 M: Add 7.4 g of calcium chloride dihydrate ($CaCl_2 \cdot 2H_2O$) to about 50 mL of distilled or deionized water and stir to dissolve. Dilute to 100 mL with water.

Iron(III) Chloride, 0.5 M: Add 13.5 g of iron(III) chloride hexahydrate ($FeCl_3 \cdot 6H_2O$) to about 50 mL of distilled or deionized water and stir to dissolve. Dilute to 100 mL with water.

Magnesium Chloride, 0.5 M: Add 10.1 g of magnesium chloride hexahydrate ($MgCl_2 \cdot 6H_2O$) to about 50 mL of distilled or deionized water and stir to dissolve. Dilute to 100 mL with water.

Sodium Hydroxide, 6 M: Cool about 50 mL of distilled or deionized water in an ice bath. (Use only borosilicate glass or a Pyrex® beaker.) Carefully add 24 g of sodium hydroxide pellets and stir to dissolve. Remove the flask from the ice bath and allow the solution to return to room temperature. Dilute to 100 mL with water.

Safety Precautions

Sodium hydroxide solution is extremely corrosive. It can cause severe skin burns and is especially dangerous to the eyes. Keep base neutralizer on hand to clean up any spills. Avoid contact of all chemicals with eyes and skin. Wear chemical splash goggles, chemical-resistant gloves, and a chemical-resistant apron. Please review relevant Material Safety Data Sheets before beginning this activity. Remind students to wash hands thoroughly with soap and water before leaving the laboratory.

Disposal

Please consult your current *Flinn Scientific Catalog/Reference Manual* for general guidelines and specific procedures governing the disposal of laboratory waste. Check the pH and appearance of the soaps if students will be allowed to take their soaps home with them.

Teacher's Notes

Lab Hints

- For best results, schedule at least two 50-minute lab periods for completion of this activity. If the experiment will be done over a 2-day period, the most convenient stopping point is after step 9. The *Pre-Lab Questions* may be assigned as homework in preparation for lab and should be reviewed before beginning the lab.

- Remind students to be very careful when adding the solid fat into the beaker. Some fat is almost always left on the outside of the beaker, which makes the beaker very slippery when heated. Fat melts at about 35 °C—heat gently and do not overheat. Handle the beaker very carefully when removing it from the heat to avoid spilling any hot oil.

- Homemade soaps may be tinted and scented by adding a few drops of food dye and perfume or cologne while stirring the soap solution and before pouring it. Many students enjoy bringing in a favorite scent from home to create a personalized soap. If students have not followed directions carefully, the soap may contain excess sodium hydroxide solution and thus be irritating to the skin. Use your judgment in deciding whether students should be allowed to take their soaps home. Alternatively, you may want to check the properties of the homemade soaps and keep them by the sink for lab use only.

- The ability of surfactants to form micelles in water can be tested by measuring their effect on the surface tension of water. The critical concentration of soap that reduces the surface tension of water is generally very low.

- This experiment offers many opportunities for a cooperative class project to investigate how the nature of fats and oils will affect the characteristics and quality of soap. Tallow is known to give a hard soap that cleans well but does not give a good lather. Oils give good sudsy lathers and also act as emollients (moisturizers). Frequently used oils and their properties include castor oil, which gives a rich and soft conditioning soap; coconut oil, which gives a creamy, fluffy lather, even in cold water; olive oil, which gives a gentle soap that is good for sensitive skin; palm kernel oil, which gives a smooth textured soap. Some oils, such as avocado oil, contain proteins and vitamins that act as natural preservatives or antioxidants in soap.

Teaching Tip

- See the *Supplementary Information* section for information about the composition of naturally occurring fats and oil and also the structures of natural fatty acids. Compare the composition of fats and oils with what is known about their health effects.

Answers to Pre-Lab Questions *(Student answers will vary.)*

1. Define the following terms: Triglyceride, saponification, surfactant, micelle.

    Triglyceride: *A compound formed from glycerol and three fatty acids. It is a triester—the fatty acids are attached via ester linkages to the glycerol backbone. Fats and oils are natural triglycerides.*

    Saponification: *Reaction of a fat or oil with sodium or potassium hydroxide. The strong base hydrolyzes or splits apart the ester linkages in the fat or oil to produce sodium or potassium salts of fatty acids (called soaps). Glycerol is a by-product of the saponification reaction.*

Teacher's Notes

Teacher Notes

Surfactant: *Also called a surface-active agent. A substance that lowers the surface tension of water and acts as an emulsifying agent. Surfactants are used as soaps and detergents.*

Micelle: *A spherical aggregate of soap molecules in aqueous solution. A micelle has two opposing features. The hydrophobic "tails" of soap molecules are arranged inward, facing each other. The exterior surface of the micelle is made up of hydrophilic ionic groups.*

2. What is the principal safety hazard in this experiment? Explain why soap made by primitive methods was likely to be very harsh.

   The manufacture of soap requires a strong base, sodium hydroxide. Sodium hydroxide solution is caustic and corrosive and can cause severe skin burns. When soap was made using primitive methods, the starting materials were not pure and the chemistry was not well understood. As a result, it was difficult to know exactly how much strong base (potash) was needed to make the soap. The soap was therefore likely to contain excess base and to be caustic.

3. Olive oil has a *saponification value* of 190 mg KOH per gram. This means that it takes 190 mg of potassium hydroxide to react completely with 1 g of olive oil.

   (a) What is the mole ratio for the reaction of a triglyceride with KOH? (See Equation 1.)

   One mole of triglyceride reacts with three moles of potassium hydroxide.

   (b) Divide 0.190 g of KOH by its molar mass to calculate the number of moles of KOH that will react with 1 g of olive oil.

   $$\frac{0.190 \text{ g KOH}}{56 \text{ g/mole}} = 0.00339 \text{ moles KOH}$$

   (c) Use the answers to (a) and (b) to determine the number of moles corresponding to 1 g of olive oil. Divide 1 g by this number of moles to calculate the average molar mass (g/mole) of olive oil.

   $$\frac{0.00339 \text{ moles KOH}}{3 \text{ moles KOH}} \times 1 \text{ mole olive oil} = 0.00113 \text{ mole olive oil}$$

   $$\frac{1 \text{ g olive oil}}{0.00113 \text{ moles}} = 885 \text{ g/mole}$$

Teacher's Notes

Teacher Notes

Sample Data

Student data will vary.

Data Table

Color, Texture and Appearance of Soap	Soap is white, very hard, and has a smooth, slippery surface.
pH of Soap (after one week)	The dried soap has a pH of 9–10.
pH and Appearance of Soap Solution	The soap solution is cloudy white and has lots of bubbles (lather). The pH of the solution is 10.
Hard Water Tests — Ca^{2+} Ions	Clumpy white precipitate suspended in solution—does not settle.
Hard Water Tests — Fe^{3+} Ions	Cloudy orange-brown precipitate forms, settles to bottom of solution.
Hard Water Tests — Mg^{2+} Ions	White precipitate settles to bottom of solution.
Emulsification Test	Lots of bubbles, cloudy white suspension with soap. Mixture of oil and water separates into two layers.

Answers to Post-Lab Questions *(Student answers will vary.)*

1. Compare the color, texture, and appearance of the homemade soap versus your favorite brand of hand soap.

   The homemade soap is similar in appearance to commercial soap. However, the surface of the homemade soap is slightly dry. Commercial soaps come in all colors.

2. Is the homemade soap solution acidic or basic? Explain.

   The soap solution is basic—pH 10.5. A commercial soap solution had a pH of 9.5.

3. How do the reactions of the soap solution with calcium, iron, and magnesium ions mimic what happens when soap is used in hard water? Draw the structure for the product obtained in the reaction of soap with calcium ions. Describe the most common result seen around the home due to the reaction of soap with hard water.

   Hard water contains high concentrations of calcium and magnesium ions. One of the problems with using soap in hard water is that the soap will not lather well and may form solid precipitates with calcium and magnesium ions. The most common result of using soap with hard water is "soap scum" left in sinks and tubs.

   $$2CH_3(CH_2)_{16}CO_2^-Na^+(aq) + Ca^{2+}(aq) \rightarrow [CH_3(CH_2)_{16}CO_2]_2Ca(s) + 2Na^+(aq)$$
   $$\text{"soap scum"}$$

Teacher's Notes

Teacher Notes

4. Compare the results of the emulsification test of olive oil with water and with soap solution. Explain in terms of the ability of soap to form micelles.

   The soap solution formed a stable emulsion with olive oil. The soap and oil emulsion appeared to be a single liquid layer and was cloudy. No oil droplets were visible. The emulsion did not separate into two layers. A mixture of water and oil separated into two layers immediately after being shaken. Soap molecules form micelles, which are able to dissolve or entrap oil molecules within their hydrophobic core.

5. Explain why most soaps contain glycerol and describe the properties that glycerol adds to soap.

   Glycerol is a by-product of the soap-making process. Unless it is specifically removed from the mixture by adding a solvent, the glycerol remains in the soap. Glycerol is a viscous liquid and is an excellent moisturizer, forming a layer that keeps moisture in.

6. Draw the general structure of a triglyceride containing both saturated and unsaturated fatty acids. Circle and label the glycerol backbone, an ester functional group, and an unsaturated fatty acid residue.

   $$\underbrace{\begin{array}{l} CH_2-O-\overset{O}{\underset{\|}{C}}-(CH_2)_{16}CH_3 \\ \,|\, \\ CH-O-\overset{O}{\underset{\|}{C}}-(CH_2)_{12}CH_3 \\ \,|\, \\ CH_2-O-\overset{O}{\underset{\|}{C}}-(CH_2)_7CH=CH(CH_2)_7CH_3 \end{array}}_{\text{Glycerol backbone}}$$

   (Ester functional group circled on top C=O; unsaturated fatty acid residue circled on bottom chain)

7. Most homemade recipes for making soap use about 5% excess fat. Explain the benefits of using excess fat to make soap.

   Using excess fat has two main benefits in soap-making. (1) Excess fat helps control the pH of soap. If there is excess fat, then all the sodium hydroxide that was added was used up in the reaction. Without excess base, the soap will not be too harsh (irritating to the skin). (2) Excess fat gives soap a smooth feel. Too much fat, however, will leave the soap greasy.

Making Soap

Teacher's Notes

Supplementary Information

Table 1. Natural Fatty Acids

Name	Structure	Classification
Lauric Acid	$CH_3(CH_2)_{10}CO_2H$	Saturated
Myristic Acid	$CH_3(CH_2)_{12}CO_2H$	Saturated
Palmitic Acid	$CH_3(CH_2)_{14}CO_2H$	Saturated
Stearic Acid	$CH_3(CH_2)_{16}CO_2H$	Saturated
Oleic Acid	$CH_3(CH_2)_7CH=CH(CH_2)_7CO_2H$	Unsaturated
Linoleic Acid	$CH_3(CH_2)_4CH=CHCH_2CH=CH(CH_2)_7CO_2H$	Polyunsaturated
Linolenic Acid	$CH_3CH_2CH=CHCH_2CH=CHCH_2CH=CH(CH_2)_7CO_2H$	Polyunsaturated

Table 2. Composition of Triglycerides—Naturally Occurring Fats and Oils

Triglyceride	Lauric	Myristic	Palmitic	Stearic	Oleic	Linoleic	Linolenic
Lard	–	1.3	28.3	11.9	47.5	6.0	–
Tallow (beef)	–	6.3	27.4	14.1	49.6	2.5	–
Coconut Oil	45.4	18.0	10.5	2.3.	7.5	Trace	–
Corn Oil	–	1.4	10.2	3.0	49.6	34.3	–
Linseed Oil	–	–	6.3	2.5	19.0	24.1	47.4
Olive Oil	–	Trace	6.9	2.3	84.4	4.6	–
Palm Oil	–	1.4	40.1	5.5	42.7	10.3	–
Safflower Oil	6.8% saturated fatty acids				18.6	70.1	3.4

Percent Composition (g Fatty Acid/100 g Total Fatty Acids)*

*Other fatty acids that may be present in amounts less than 1% include C_6, C_8, C_{10} and C_{20} saturated; C_{12}, C_{14}, and C_{16} monounsaturated; and C_{20} polyunsaturated.

Page 1 – **Preparation of Esters**

Teacher Notes

Preparation of Esters
Nature's Flavors and Fragrances

Introduction

What do you "taste" when you bite into an apple or a banana? The unique flavor of any food is due to a combined sense of both taste and smell. Indeed, the first taste perception of any food comes from the aroma or fragrance of volatile organic compounds. In the case of fruits, the primary flavor and fragrance ingredients are organic compounds called esters. What are esters and how can they be prepared in the lab?

Concepts

- Ester functional group
- Carboxylic acids and alcohols
- Esterification reaction
- Equilibrium

Background

The study of organic chemistry is organized around *functional groups*—groups of atoms, bonded together in a specific pattern, that give organic compounds their unique physical and chemical properties. For example, compounds belonging to the same functional group class show similar trends in their solubility and boiling points and also undergo characteristic types of chemical reactions.

The structure of the *ester* functional group is shown in Figure 1. The "R" groups represent any combination of carbon and hydrogen atoms and are called *alkyl groups*. The C and H atoms in the alkyl groups may be bonded together to form either chain or ring structures.

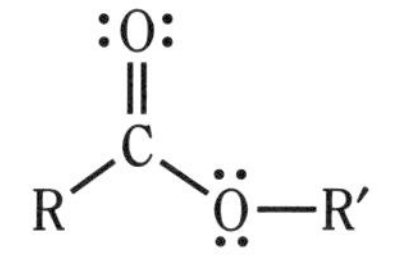

Figure 1. Structure of the Ester Functional Group.

Organic esters are widely distributed in nature. Low molecular weight esters are responsible for the pleasant odor or fragrance of many fruits and flowers, and they are important ingredients in natural and artificial flavors. More than 100 esters (and all of the esters that will be made in this lab) are designated by the FDA as GRAS—*generally regarded as safe*. This means that the esters may be used as food additives *without* going through a testing and approval process. Natural banana flavor, for example, is due primarily to four esters (Figure 2). Notice the ester functional groups, which are circled, and the variety of different alkyl groups attached to the ester linkage.

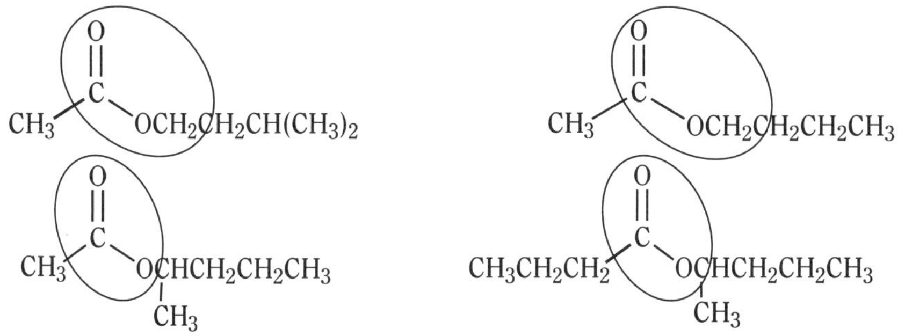

Figure 2. Ester Components of Natural Banana Flavor.

The Food and Drug Administration (FDA) is responsible for ensuring that the food we eat is safe and wholesome and that the medicines we use are safe and effective. The FDA reviews all test results for companies seeking approval of food additives, drugs, and vaccines. The agency also guarantees that these substances are truthfully labeled.

Preparation of Esters – Page 2

Esters are considered derivatives of carboxylic acids—they may be prepared by the reaction of a carboxylic acid with an alcohol. The preparation of ethyl acetate illustrates the general principles of ester synthesis. Ethyl acetate is an important industrial solvent. It is used in many consumer items, such as nail polish, and it is also a naturally occurring "flavor ingredient" in apples and bananas. Ethyl acetate is obtained by heating a solution of ethyl alcohol and acetic acid in the presence of a strong acid catalyst such as sulfuric acid (Equation 1). The synthesis of esters demonstrates the usefulness of the functional group concept in organic chemistry. With few exceptions and under the proper conditions, most carboxylic acids and alcohols will undergo this type of esterification reaction.

$$CH_3-\overset{O}{\underset{\|}{C}}-O-H + CH_3-CH_2-O-H \underset{}{\overset{H_2SO_4}{\rightleftarrows}} CH_3-\overset{O}{\underset{\|}{C}}-O-CH_2-CH_3 + H_2O \qquad \text{Equation 1}$$

Acetic acid *Ethyl alcohol* *Ethyl acetate* *Water*

The use of the double arrow in Equation 1 indicates that ester synthesis is reversible. The reverse reaction of water with ethyl acetate splits apart the ester functional group and is called *hydrolysis*. Under typical conditions, with heat and an acid catalyst, the forward and the reverse reactions will quickly reach equilibrium. Both reactants and products may be present in significant amounts at equilibrium, limiting the yield of ester that may be obtained. The equilibrium constant for the formation of ethyl acetate illustrates this problem. At 25 °C, the value of the equilibrium constant for the synthesis of ethyl acetate is approximately four (Equation 2). Starting with one mole of ethyl alcohol and one mole of acetic acid, the maximum amount of ester at equilibrium will be 0.67 mole (67% yield). The yield of ester may be increased by shifting the equilibrium to the right using LeChátelier's principle—by using excess ethyl alcohol, for example, or by removing the water from the reaction mixture as it forms.

$$K_{eq} = \frac{[CH_3CO_2CH_2CH_3][H_2O]}{[CH_3CO_2H][CH_3CH_2OH]} = 4 \qquad \text{Equation 2}$$

Esters are named as derivatives of an alcohol and carboxylic acid. The name of an ester is always two words, where the first word comes from the name of the alcohol, the second word from the name of the carboxylic acid by changing the *-ic* ending to *-ate*. The examples in Table 1 illustrate the names and formulas of some naturally occurring esters.

Table 1. Naturally Occurring Esters

Natural Flavor	Name of Ester	Formula of Ester
Oranges	Octyl acetate	$CH_3-\overset{O}{\underset{\|}{C}}-OCH_2CH_2CH_2CH_2CH_2CH_2CH_3$
Apples	Methyl butyrate	$CH_3CH_2CH_2-\overset{O}{\underset{\|}{C}}-OCH_3$
Pineapple	Ethyl butyrate	$CH_3CH_2CH_2-\overset{O}{\underset{\|}{C}}-O-CH_2CH_3$

Page 3 – **Preparation of Esters**

Teacher Notes

Experiment Overview

The purpose of this experiment is to investigate the preparation of organic esters starting with alcohols and carboxylic acids. The esters will be identified by their characteristic odors.

Pre-Lab Questions

1. Ester formation is a very versatile reaction—many different carboxylic acids and alcohols may be used. In this experiment, five alcohols and four carboxylic acids will be available as starting materials (Table 2). Fill in the blanks with the name of the ester that would be obtained from each combination of an alcohol and a carboxylic acid. The first box has been completed for you. *Note:* The combinations that are shaded will not be used and may be left blank.

Table 2. Preparation of Esters from Alcohols and Carboxylic Acids

	Acetic Acid	Propionic Acid	Benzoic Acid	Salicylic Acid
Methyl Alcohol				
Ethyl Alcohol	Ethyl acetate			
Propyl Alcohol				
Isoamyl Alcohol				
Octyl Alcohol				

2. Draw the structures of (a) ethyl propionate and (b) methyl benzoate. See the *Materials* section for the formulas of the alcohol and acid precursors of these esters.

3. Write a balanced chemical equation for the preparation of (a) octyl acetate and (b) methyl salicylate from the necessary starting materials.

4. What is the role of concentrated sulfuric acid in ester synthesis? What precautions are necessary when working with concentrated sulfuric acid in the lab?

Preparation of Esters – Page 4

Materials

Alcohols (4 mL each)
Ethyl alcohol, CH_3CH_2OH
Isoamyl alcohol, $(CH_3)_2CHCH_2CH_2OH$
Methyl alcohol, CH_3OH
Octyl alcohol, $CH_3(CH_2)_6CH_2OH$
Propyl alcohol, $CH_3CH_2CH_2OH$

Carboxylic acids (4 mL each)
Acetic acid, CH_3CO_2H
Benzoic acid, $C_6H_5CO_2H$*
Propionic acid, $CH_3CH_2CO_2H$
Salicylic acid, $HOC_6H_4CO_2H$†
Sulfuric acid, H_2SO_4, concentrated, 2 mL

Beaker, 400-mL
Hot plate
Beral pipets, 4
Pasteur pipet, glass
Distilled water and wash bottle
Graduated cylinders, 10 mL, 2
Sodium bicarbonate solution, $NaHCO_3$ saturated, 8 mL
Test tube clamp
Test tube rack
Test tubes, medium, 8
Thermometer
Watch glasses, 4
Wax marking pencil

*Benzoic acid

†Salicylic acid

Safety Precautions

Concentrated sulfuric acid is severely corrosive to eyes, skin, and body tissues. Notify the teacher and clean up all spills, even a few drops, immediately! Acetic acid is corrosive to skin and body tissue. It is a moderate fire risk and is toxic by ingestion and inhalation. Methyl alcohol is extremely flammable and a dangerous fire risk. It is toxic by ingestion and may cause blindness. Ethyl alcohol is a flammable solvent and toxic by ingestion. Propionic acid is a flammable liquid and irritating to skin and eyes. It is slightly toxic by ingestion and has a rancid odor. Salicylic acid is moderately toxic by ingestion. Propyl alcohol is irritating to skin and eyes and is slightly toxic by ingestion. Benzoic acid is slightly toxic by ingestion and is a body tissue irritant. Avoid contact of all chemicals with skin and eyes. Do not use any flames in the laboratory when working with alcohols and other flammable solvents. Keep away from all sources of ignition.

Perform this experiment in a well-ventilated lab only and avoid inhaling the vapors. To smell a product, carefully waft the vapors to your nose—do NOT "sniff"! Wear chemical splash goggles, chemical-resistant gloves, and a chemical-resistant apron. Wash hands thoroughly with soap and water before leaving the laboratory.

Teacher Notes

Please do not be intimidated by the comprehensive safety warnings included for this lab. Carrying out the experiment on the mini-scale level described in this experiment minimizes the risks associated with working with these chemicals. Some of the reagents are flammable—do not use any flames. Other reagents are corrosive—wear goggles, gloves, and apron!

Page 5 – **Preparation of Esters**

Teacher Notes

Procedure

1. Fill a 400-mL beaker two-thirds full with hot tap water. Heat the water to about 80 °C on a hot plate.

2. Choose two alcohols and two carboxylic acids from the list of materials. Write the names of the alcohols and acids selected for testing in the data table. *Note:* In order to test all possible combinations, the teacher may select the alcohols and carboxylic acids that will be tested by each group.

3. Obtain four medium test tubes. Make sure each test tube is clean *and* dry. Label the test tubes with the name or an abbreviation for each reagent selected (e.g., MeOH for methyl alcohol, BA for benzoic acid, etc.).

4. Obtain about 4 mL of each alcohol and acid to be tested in the corresponding test tube. *Note:* In the case of solids, obtain about 2 grams of each compound.

5. Label a second set of four test tubes #1–4. In the data table, write in the names of the alcohol and the carboxylic acid that will be combined in each test tube. *Example:* Methyl alcohol and benzoic acid in test tube #1, methyl alcohol and salicylic acid in test tube #2, etc.

6. Using a clean graduated Beral pipet or a spatula, transfer 2 mL of the appropriate alcohol and 2 mL (or 1 g for a solid) of the appropriate carboxylic acid into test tube #1. *Example:* Place 2 mL of methyl alcohol and 1 g of benzoic acid into test tube #1.

7. Repeat step 6 for test tubes #2, 3, and 4. Place all of the test tubes in a test tube rack.

8. Take the test tube rack to a central location where the concentrated sulfuric acid is being dispensed. Using a glass eyedropper or Pasteur pipet, carefully add 10 drops of concentrated sulfuric acid to each test tube #1–4. *Caution:* Exercise extreme care when working with concentrated sulfuric acid.

9. Return the test tube rack to the lab bench and place each test tube #1–4 into the hot water bath at 80 °C (step 1).

10. Heat the reaction mixtures in test tubes #1–4 in the hot water bath for about 10 minutes. Record any changes in the appearance of the materials in the data table.

11. After 10 minutes, use a test tube clamp to remove the test tubes from the hot water bath. Replace the test tubes in the test tube rack and allow them to cool for about 5 minutes. Record any observations in the data table.

12. Add about 2 mL of saturated sodium bicarbonate solution to each test tube. Record observations in the data table.

13. Using a glass Pasteur pipet, remove 2–3 drops of liquid from the upper layer in test tube #1. Place the drops on a clean watch glass and carefully *waft* the vapors from the watch glass to your nose to smell the product. See Figure 3.

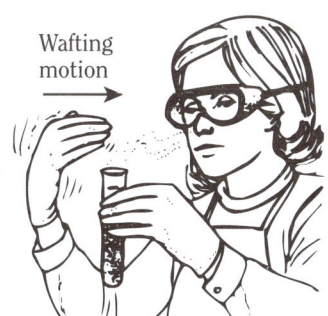

Figure 3.

14. In the data table, describe the odor of the ester product. *Hint:* Some odors will be easy to identify. Adjectives that fragrance chemists use to describe odors include floral, fruity, earthy, pungent, musky, sweet, herbal, green, etc.

15. Dispose of the contents of the test tube according to your instructor's directions.

If using octyl alcohol as one of the reagents, add only 1 mL of the alcohol, not 2 mL. Excess octyl alcohol will give the ester a biting, medicinal smell.

Preparation of Esters

Preparation of Esters – Page 6

Name: _____

Class/Lab Period: _____

Preparation of Esters

Data Table

Alcohols to be tested	
Carboxylic acids to be tested	

Test Tube	Alcohol	Carboxylic Acid	Observations of Ester and Odor
1			
2			
3			
4			

Post-Lab Questions *(Use a separate sheet of paper to answer the following questions.)*

1. Write a chemical equation for the formation of each ester in test tubes #1–4 and write the name of each ester product next to its structure.

2. Describe in general terms the odor or fragrance of the esters that were prepared by you and others in the class. Compare the odors of the esters to the odors of the starting materials.

3. Were any of the esters easily identified as a specific fragrance, e.g., apple or banana? In cases where a specific fragrance was detected, how does the odor compare to the natural fragrance? Give one reason for any difference between the synthetic fragrance and the natural fragrance.

4. Sodium bicarbonate solution was added to the product mixtures in step 12 to remove any unreacted acid. Write a balanced chemical equation for the reaction of acetic acid with sodium bicarbonate.

5. Ethyl propionate and propyl acetate are isomers. Consider the molecular formula and the structures of these two compounds and write a definition of isomers based on this comparison.

Teacher's Notes
Preparation of Esters

Master Materials List *(for a class of 30 students working in pairs)*

Alcohols (60 mL each)
Ethyl alcohol, CH_3CH_2OH
Isoamyl alcohol, $(CH_3)_2CHCH_2CH_2OH$
Methyl alcohol, CH_3OH
Octyl alcohol, $CH_3(CH_2)_6CH_2OH$
Propyl alcohol, $CH_3CH_2CH_2OH$

Carboxylic acids (60 mL each)
Acetic acid, CH_3CO_2H
Benzoic acid, $C_6H_5CO_2H$, 30 g*
Propionic acid, $CH_3CH_2CO_2H$
Salicylic acid, $HOC_6H_4CO_2H$, 30 g†

Sulfuric acid, H_2SO_4, concentrated, 30 mL
Beakers, 400-mL, 5–7‡
Hot plates, 5–7‡
Beral pipets, 60
Pasteur pipets, glass, 15
Distilled water and wash bottles, 15
Graduated cylinders, 10 mL, 30
Test tubes, medium, 120
Test tube racks, 15
Watch glasses, 60
Wax marking pencils, 15

*Benzoic acid

†Salicylic acid.

‡*Several groups may share hot plates and hot water baths.*

Safety Precautions

Perform this experiment in a well-ventilated lab only and avoid inhaling any vapors. Concentrated sulfuric acid is severely corrosive to eyes, skin, and body tissues. Keep sodium carbonate or another neutralizing agent on hand to clean up all spills immediately. Acetic acid is corrosive to skin and body tissue. It is a moderate fire risk and is toxic by ingestion and inhalation. Methyl alcohol is extremely flammable and a dangerous fire risk. It is toxic by ingestion and may cause blindness. Ethyl alcohol is a flammable solvent and toxic by ingestion. Propionic acid is a flammable liquid and is irritating to skin and eyes. It is slightly toxic by ingestion and has a rancid odor. Salicylic acid is moderately toxic by ingestion. Propyl alcohol is irritating to skin and eyes and is slightly toxic by ingestion. Benzoic acid is slightly toxic by ingestion and is a body tissue irritant. Avoid contact of all chemicals with skin and eyes. Do not use any flames in the laboratory when working with alcohols and other flammable solvents. Keep away from all sources of ignition.

Instruct students in the proper procedure for smelling chemical odors—carefully waft the vapors to the nose—do NOT "sniff"! Wear chemical splash goggles, chemical-resistant gloves, and a chemical-resistant apron. Please review current Material Safety Data Sheets for additional safety, handling, and disposal information. Remind students to wash their hands thoroughly with soap and water before leaving the laboratory.

Teacher's Notes

Disposal

Please consult your current *Flinn Scientific Catalog/Reference Manual* for general guidelines and specific procedures governing the disposal of laboratory waste. All aqueous solutions may be disposed of down the drain with plenty of excess water according to Flinn Suggested Disposal Method #26b. Ester products may be disposed of using Flinn Suggested Disposal Method #18a.

Lab Hints

- The laboratory work for this experiment can easily be completed in a typical 50-minute lab period. Divide up the possible alcohol and carboxylic acid combinations among different student groups so that all of the possible esters shown in Table 2 may be observed. To avoid sensory overload in the lab from too many different products, the number of esters to be investigated was reduced from the 20 possible combinations to 12. All of the possible acid and alcohol combinations shown in Table 2 will work and the esters, with the exception of octyl benzoate and octyl salicylate, are biodegradable, have pleasant odors, and are "generally regarded as safe" (GRAS) by the FDA.

- Although many butyrate and valerate esters have pleasant and easily identifiable odors, we chose not to use either butyric (C_4) or valeric acid (C_5) in this lab. Both acids emit a piercing stench—the odor may linger for weeks.

- Concentrated sulfuric acid is severely corrosive. For best results, dispense concentrated sulfuric acid *on a spill tray* in the hood and *monitor the students* carefully as they work with the acid.

- Methyl alcohol is extremely flammable and a severe fire risk. Keep away from heat, flames, and other sources of ignition. Do NOT under any circumstances use Bunsen burners to heat the water baths. Teachers on a budget who cannot afford hot plates have told us that they purchase old percolators or coffee makers to provide a source of hot water for the lab.

- Perform this experiment in a well-ventilated lab only. With so many different esters being prepared, it may be difficult for students to differentiate and identify the various odors. Keep a box of unscented tissues in the lab for students to cleanse their noses before smelling a new compound. If ventilation is a problem, you may want to choose one ester and have all students synthesize the same product. Methyl salicylate and isoamyl acetate are easily identifiable and good choices.

Teaching Tip

- The most commonly used esters for artificial flavors are: ethyl butyrate, methyl salicylate, isoamyl acetate, ethyl propionate, ethyl anthranilate, butyl butyrate, and isobutyl acetate. Esters are also widely used as fragrances to make perfumes. The following Web site is a good source of information about the use of esters: http://www.thegoodscentscompany.com. Select "Perfumery Raw Materials" and scroll through the chemical names to locate a specific ester. Click on the name of the ester to obtain odor descriptions and information about the natural occurrence, physical properties, and perfumery uses. Even though most of the esters prepared in this lab are classified as GRAS, they must still be treated as potentially harmful laboratory chemicals. Do not allow ester products to be removed from the lab.

For a list of GRAS additives, please consult the following Web site: http://vm.cfsan.fda.gov/%7Edms/eafus.html

Teacher's Notes

Teacher Notes

Answers to Pre-Lab Questions *(Student answers will vary.)*

1. Ester formation is a very versatile reaction—many different carboxylic acids and alcohols may be used. In this experiment, five alcohols and four carboxylic acids will be available as starting materials (Table 2). Fill in the blanks with the name of the ester that would be obtained from each combination of an alcohol and a carboxylic acid. The first box has been completed for you. *Note:* The combinations that are shaded will not be used and may be left blank.

Table 2. Preparation of Esters from Alcohols and Carboxylic Acids

	Acetic Acid	Propionic Acid	Benzoic Acid	Salicylic Acid
Methyl Alcohol			Methyl benzoate	Methyl salicylate
Ethyl Alcohol	Ethyl acetate	Ethyl propionate	Ethyl benzoate	Ethyl salicylate
Propyl Alcohol	Propyl acetate	Propyl propionate		
Isoamyl Alcohol	Isoamyl acetate	Isoamyl propionate		
Octyl Alcohol	Octyl acetate	Octyl propionate		

2. Draw the structures of (a) ethyl propionate and (b) methyl benzoate. See the *Materials* section for the formulas of the alcohol and acid precursors of these esters.

   (a) Ethyl propionate $\quad CH_3CH_2-\overset{\overset{O}{\|}}{C}-O-CH_2CH_3$

   (b) Methyl benzoate

3. Write a balanced chemical equation for the preparation of (a) octyl acetate and (b) methyl salicylate from the necessary starting materials.

   (a) $CH_3\overset{\overset{O}{\|}}{C}OH + CH_3(CH_2)_7OH \rightleftharpoons CH_3\overset{\overset{O}{\|}}{C}-O(CH_2)_7CH_3$

   (b) salicylic acid + $CH_3OH \rightleftharpoons$ methyl salicylate

4. What is the role of concentrated sulfuric acid in ester synthesis? What precautions are necessary when working with concentrated sulfuric acid in the lab?

   Sulfuric acid is used as an acid catalyst to speed up the rate of ester formation. Sulfuric acid is severely corrosive. Exercise extreme care when using sulfuric acid and work with sulfuric acid only in a designated area. Notify the teacher immediately if any acid is spilled.

Teacher's Notes

Sample Data

Student data will vary.

Data Table

Alcohols to be tested	Isoamyl alcohol and octyl alcohol		
Carboxylic acids to be tested	Acetic acid and propionic acid		
Test Tube	**Alcohol**	**Carboxylic Acid**	**Observations of Ester and Odor**
1	Isoamyl alcohol	Acetic acid	The upper layer is colorless and oily. Very distinct banana smell.
2	Isoamyl alcohol	Propionic acid	Upper layer is pale pink color, oily. Fruity, sweet peach smell.
3	Octyl alcohol	Acetic acid	The upper layer is yellow and oily. Fruity, citrus smell, slightly medicinal.
4	Octyl alcohol	Propionic acid	Upper layer is pale yellow and oily. Sweet smell, but unpleasant.

Answers to Post-Lab Questions *(Student answers will vary.)*

1. Write a chemical equation for the formation of each ester in test tubes #1–4 and write the name of each ester product next to its structure.

$$CH_3CO_2H + (CH_3)_2CHCH_2CH_2OH \rightleftharpoons CH_3\overset{O}{\overset{\|}{C}}OCH_2CH_2CH(CH_3)_2 + H_2O$$
Isoamyl acetate

$$CH_3CH_2CO_2H + (CH_3)_2CHCH_2CH_2OH \rightleftharpoons CH_3CH_2\overset{O}{\overset{\|}{C}}OCH_2CH_2CH(CH_3)_2 + H_2O$$
Isoamyl propionate

$$CH_3CO_2H + CH_3(CH_2)_6CH_2OH \rightleftharpoons CH_3\overset{O}{\overset{\|}{C}}OCH_2(CH_2)_6CH_3 + H_2O$$
Octyl acetate

$$CH_3CH_2CO_2H + CH_3(CH_2)_6CH_2OH \rightleftharpoons CH_3CH_2\overset{O}{\overset{\|}{C}}OCH_2(CH_2)_6CH_3 + H_2O$$
Octyl propionate

Teacher's Notes

Teacher Notes

2. Describe in general terms the odor or fragrance of the esters that were prepared by you and others in the class. Compare the odors of the esters to the odors of the starting materials.

   Most of the esters have pleasant smells. The alcohols have medicinal smells, while the acids are sharp, biting, and unpleasant. The odors of all of the possible ester products are described here:

Methyl benzoate	*Minty, dull, medicinal*
Methyl salicylate	*Distinct wintergreen smell*
Ethyl acetate	*Sweet odor, nail polish remover*
Ethyl propionate	*Strong, fruity odor, rum, butterscotch*
Ethyl benzoate	*Sweet, pleasant, tropical, fruity*
Ethyl salicylate	*Minty, sweet*
Propyl acetate	*Fruity, pears, peach*
Propyl propionate	*Medicinal, fruity*
Isoamyl acetate	*Strong, fruity, distinct banana smell*
Isoamyl propionate	*Sweet, fruity, ripe, peach*
Octyl acetate	*Fruity, biting, pineapple*
Octyl propionate	*Fruity, musty*

3. Were any of the esters easily identified as a specific fragrance, e.g. apple or banana? In cases where a specific fragrance was detected, how does the odor compare to the natural fragrance? Give one reason for any difference between the synthetic fragrance and the natural fragrance.

   There were some very distinct odors that were easy to identify. Isoamyl acetate (banana) and methyl salicylate (wintergreen) stand out. The odors of these compounds, while very distinct, were stronger and less subtle than the natural fragrances. The natural fragrances are more of a blend of several different odors. Natural fragrances are made up of many different compounds. Apple flavor, for example, is a blend of at least 29 volatile compounds.

4. Sodium bicarbonate solution was added to the product mixtures in step 12 to remove any unreacted acid. Write a balanced chemical equation for the reaction of acetic acid with sodium bicarbonate.

   $$CH_3CO_2H(aq) + NaHCO_3(aq) \rightarrow CH_3CO_2Na(aq) + H_2O(l) + CO_2(g)$$

5. Ethyl propionate and propyl acetate are isomers. Consider the molecular formula and the structures of these two compounds and write a definition of isomers based on this comparison.

   Ethyl propionate and propyl acetate have the same molecular formula—$C_5H_{10}O_2$. They have different structures, however. Isomers are compounds that have the same formula but different structures. Isomers are different compounds and have different physical properties.

Teacher's Notes

Teacher Notes

Synthesis of Aspirin
From Natural Products to Painkillers

Introduction

Aspirin, first synthesized in 1897, is one of the oldest yet most common drugs in use today. Like many modern drugs, aspirin has its roots in an ancient folk remedy—the use of willow extracts to treat fever and pain. Aspirin is prepared the same way today that it was more than 100 years ago. Let's look at the structure, synthesis, and properties of aspirin.

Concepts

- History of aspirin
- Esters and esterification
- Salicylic acid derivatives
- Excess and limiting reagents

Background

Native Americans, as well as the ancient Chinese, Egyptians, and Greeks, used willow extracts to treat fever, pain, and inflammation. The Ebers papyrus, dating to at least 1500 B.C. in Egypt, contains the earliest written reference to the use of willow extracts, "to draw the heat out" from inflammation. Willow extracts remained a popular folk medicine remedy throughout the Middle Ages. The first scientific study of the effectiveness of willow extracts was carried out in 1763 by the Rev. Edward Stone in England. In one of the first ever "clinical trials" of a drug, Stone reported using willow extracts to treat fever and pain in more than 50 patients suffering from malaria.

In the early 19th century, organic chemistry was only a fledgling science, with roots in the study of natural products. In 1828, Johann Büchner at the University of Munich in Germany isolated a crystalline compound from willow bark and named it *salicin,* after the Latin name for the white willow, *Salix alba.* Ten years later, the Italian chemist Raffaele Piria converted salicin to salicylic acid, which had also recently been isolated from meadowsweet (*Spiraea*). *Salicylic acid* (Figure 1) was found to be the active ingredient responsible for the medicinal properties of many plants, including willow, poplar, aspen, and myrtle. In 1859, Hermann Kolbe at Marburg University in Germany determined the chemical structure of salicylic acid and synthesized it from phenol, a derivative of coal tar. By 1870 salicylic acid was widely used in Europe for the treatment of arthritis, pain, and fever. Unfortunately, the compound was "tough to swallow" and very irritating to the stomach. Many people could not tolerate the drug because of its severe and unpleasant side effects.

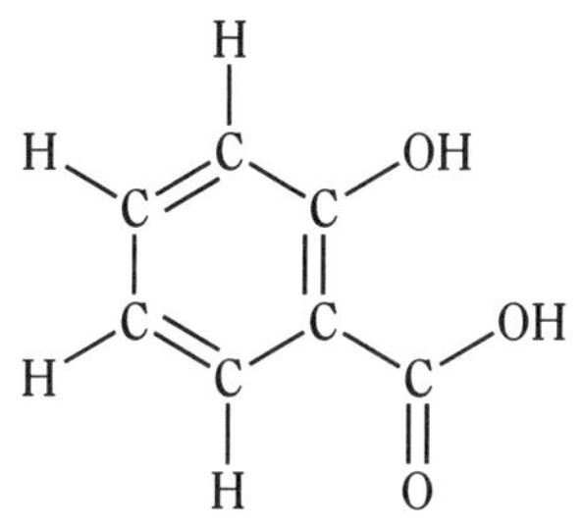

Figure 1. Structure of Salicylic Acid

Synthesis of Aspirin – Page 2

Felix Hoffmann, an organic chemist working at Friedrich Bayer and Company in Germany, attempted to chemically modify salicylic acid and thus reduce its side effects. In 1897, Hoffmann synthesized *acetylsalicylic acid* by reacting salicylic acid with acetic anhydride (Equation 1).

Equation 1

Salicylic acid   *Acetic anhydride*   *Aspirin—Acetylsalicylic acid*

The synthesis of acetylsalicylic acid is an example of an *esterification* reaction in which the phenolic –OH group in salicylic acid is replaced with an acetyl or ester functional group (–OCOCH$_3$). Masking the –OH functional group in this way makes the compound less corrosive. Acetylsalicylic acid is an effective analgesic (pain reliever) and antipyretic (fever reducer) but is less acidic or harsh than salicylic acid. In 1899, the Bayer Company marketed acetylsalicylic acid under the trade name *aspirin,* with *a-* denoting the acetyl group and *–spirin* referring to *Spiraea,* the plant from which salicylic acid was first isolated. It is estimated that approximately 50 billion aspirin tablets are consumed per year all over the world, and that as many as one *trillion* (1×10^{12}) aspirin tablets have been produced in the 100 years since its discovery!

Acetylsalicylic acid remains a versatile drug in the 21st century. The two most common uses of aspirin today are for the prevention of heart attack and stroke and to relieve the pain and reduce the inflammation of arthritis. The American Heart Association recommends "an aspirin a day" to prevent a second heart attack in individuals who have had a previous heart attack or stroke. The myriad physiological effects of aspirin were explained in 1972 by Sir John Vane (Nobel Prize in Medicine, 1982) and coworkers at the Wellcome Research Laboratories in Great Britain. Vane found that aspirin inhibited an enzyme involved in the synthesis of prostaglandins and thus interfered with their production in the body. *Prostaglandins* are hormone-like "chemical messengers" that play a key role in a variety of physiological processes, including inflammation, blood clotting, labor and childbirth, and blood pressure. Aspirin prevents the formation of blood clots that are a major cause of heart attacks and strokes.

Experiment Overview

The purpose of this experiment is to prepare acetylsalicylic acid (aspirin), determine its purity, and investigate its chemical properties.

Teacher Notes

Pre-Lab Questions

1. Acetic anhydride is a *lachrymator*. What is a lachrymator and what safety precautions should be followed when working with acetic anhydride?

2. What is concentrated sulfuric acid used for in this experiment? What are the hazards of working with concentrated sulfuric acid?

3. Calculate (a) the molar mass of salicylic acid and acetic anhydride and (b) the number of moles of each that will be used in this experiment. *Note:* The density of acetic anhydride is 1.08 g/mL.

4. Define the term *limiting reactant*. Complete the following statement: The maximum number of moles of aspirin that can be obtained in this experiment is equal to the number of moles of _____ used.

5. Determine the chemical formula of acetylsalicylic acid and calculate its molar mass.

Materials

Acetic anhydride, $(CH_3CO)_2O$, 1 mL
Aspirin tablet, crushed
Ethyl alcohol, 50%, 6 mL (optional)
Iron(III) chloride solution, $FeCl_3$, 0.1 M, 2 mL
Salicylic acid, $HO–C_6H_4–CO_2H$, 0.5 g
Sulfuric acid, concentrated, H_2SO_4, 2 drops
Distilled water and wash bottle
Ice, crushed
Balance, 0.01-g precision
Beaker, 50-mL
Beakers, 250-mL, 2
Beral-type pipets, graduated, 4
Boiling stone

Capillary tubes
Erlenmeyer flasks, 50- and 250-mL
Filter paper (to fit funnel)
Funnel
Graduated cylinder, 10-mL
Hot plate
Melting point apparatus or Thiele-Dennis tube
Pasteur pipet or eyedropper
Ring (support) stand and clamp
Stirring rod
Test tubes, small, 3
Thermometer
Watch glass

Safety Precautions

Concentrated sulfuric acid is severely corrosive to eyes, skin, and body tissue. Notify the teacher and clean up all spills immediately. Acetic anhydride is a corrosive liquid and the vapors are highly irritating. The liquid is flammable and a strong lachrymator—contact with the liquid will cause severe eye irritation. Work with acetic anhydride in the hood or in a well-ventilated lab only. Do not inhale the vapors. Salicylic acid is moderately toxic by ingestion. Avoid contact of all chemicals with eyes and skin. Wear chemical splash goggles, chemical-resistant gloves, and a chemical-resistant apron. Wash hands thoroughly with soap and water before leaving the laboratory.

Synthesis of Aspirin – Page 4

Procedure

Teacher Notes

1. Fill a 250-mL beaker about two-thirds full with hot tap water and add a boiling stone.

2. Place a hot plate on the base of a ring (support) stand and place the beaker on the hot plate. Heat the water to about 80 °C using a medium-high setting of the hot plate.

3. Tare (zero) a clean 50-mL Erlenmeyer flask. Place about 0.5 g of salicylic acid in the flask and measure the precise mass to 0.01 g. Record the mass of salicylic acid in the data table.

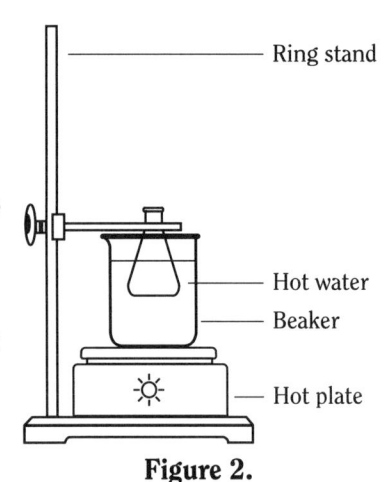

Figure 2.

4. Take the Erlenmeyer flask to the hood where the acetic anhydride is being dispensed. Add 1 mL of acetic anhydride to the flask using a graduated Beral-type pipet.

5. Using a glass eyedropper or Pasteur pipet, *carefully* add 2 drops of concentrated sulfuric acid to the Erlenmeyer flask.

6. Place the Erlenmeyer flask in a clamp and attach the clamp to the ring stand. Carefully lower the Erlenmeyer flask into the hot water bath. See Figure 2.

7. Heat the reaction mixture in the hot water bath for 10 minutes.

8. Half-fill a 250-mL beaker with crushed ice and water to use as an ice bath. Obtain about 15 mL of distilled water in a small beaker or test tube and cool the water in the ice bath.

9. After 10 minutes, *carefully* lift the Erlenmeyer flask out of the hot water bath.

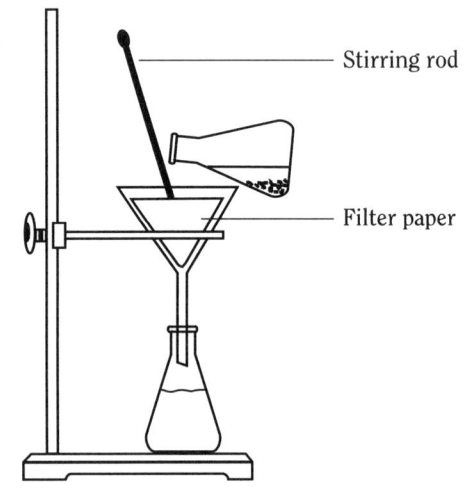

Figure 3.

10. Allow the Erlenmeyer flask to cool, then add 3 mL of ice-cold water (from step 8) dropwise to the reaction mixture. *Note:* Water may react vigorously with acetic anhydride to form acetic acid.

11. Add an additional 5–6 mL of ice-cold water to the Erlenmeyer flask and place the flask in the ice bath (step 8) to allow the aspirin to crystallize. If no crystals have formed after 5 minutes, remove the flask from the ice and scratch the sides of the flask with a stirring rod.

12. Keep the flask in the ice-water bath for 10 minutes to complete crystal formation.

13. Set up a funnel for gravity filtration as shown in Figure 3. Place a clean beaker or flask under the funnel to collect the filtrate. Wet the filter paper with a few drops of distilled water.

Teacher Notes

14. Using a stirring rod to direct the stream of liquid, slowly pour the reaction mixture from the Erlenmeyer flask into the funnel. Gently swirl the flask to get as much of the solid as possible into the funnel with just one pour.

15. When most of the liquid has passed through the funnel, rinse any remaining crystals from the flask into the funnel with a *small amount* (no more than 4 mL) of ice-cold water.

16. Measure and record the mass of a clean and dry watch glass. When there is no more liquid in the funnel, carefully remove the filter paper and scrape the crystals onto the preweighed watch glass. *Note:* If the product will be recrystallized (step 17), transfer the crystals to a clean 50-mL Erlenmeyer flask.

17. *(Optional)* To recrystallize the product, dissolve in 3 mL of ethyl alcohol and gently heat (do not boil) the mixture on a hot plate. Add about 6 mL of distilled water to the hot solution until the solution is slightly cloudy. Cool the flask in an ice bath to obtain crystals and then repeat steps 13–16 to filter and wash the crystals.

18. Label the watch glass with your initials and allow the crystals to air dry for at least 2 hours. Measure and record the combined mass of the watch glass and aspirin in the data table.

19. Label three small test tubes *A–C* and add a small amount (about 20 mg) of (a) salicylic acid, (b) the reaction product, and (c) crushed aspirin to the appropriate test tube. Add about 2 mL of 50% ethyl alcohol to each test tube to dissolve the solids.

20. Add 3 drops of 0.1 M iron(III) chloride solution to each test tube. Record observations in the data table.

21. Measure the melting point of the reaction product.

See the Lab Hints section for instructions on how to use a Thiele–Dennis tube to measure melting points.

Synthesis of Aspirin

Synthesis of Aspirin – Page 6

Name: _____

Class/Lab Period: _____

Synthesis of Aspirin

Data Table

Mass of salicylic acid used	
Mass of watch glass	
Mass of watch glass and acetylsalicylic acid	
Melting point of acetylsalicylic acid	
Results of Fe^{3+} Tests	
Salicylic acid	
Reaction product	
Crushed aspirin	

Post-Lab Questions *(Answer the following questions on a separate sheet of paper.)*

1. Calculate the number of moles of salicylic acid used in this experiment.

2. Calculate the maximum amount of acetylsalicylic acid in grams that may be obtained from this amount of salicylic acid. This is the theoretical yield. *Hint:* See *Pre-Lab Questions* #4 and 5.

3. Determine the mass of aspirin obtained in this experiment and calculate the *percent yield*.

$$\text{Percent yield} = \frac{\text{Actual yield}}{\text{Theoretical yield}} \times 100\%$$

4. Iron(III) ions are used as a qualitative test for *phenols* (aromatic compounds containing an –OH functional group). (a) What compound was used as a *positive control* for the Fe^{3+} test in this experiment? (b) Did the reaction product give a positive or negative test result with Fe^{3+} ions? Explain.

5. Old aspirin tablets often have a faint vinegar (acetic acid) smell and give a positive test with iron(III) ions. Write a balanced chemical equation for the *hydrolysis* of aspirin (reaction of aspirin with water) to explain these observations.

6. Acetic anhydride was used in excess in this experiment. What does this mean, and how was the excess acetic anhydride decomposed at the end of the reaction?

7. Look up the melting points of salicylic acid and aspirin (acetylsalicylic acid) in a reference book or online and compare with the melting point of the reaction product.

Teacher Notes

Teacher's Notes
Synthesis of Aspirin

Master Materials List *(for a class of 30 students working in pairs)*

Acetic anhydride, $(CH_3CO)_2O$, 15 mL	Capillary tubes
Aspirin tablet, crushed	Erlenmeyer flasks, 50-mL, 15
Ethyl alcohol, 50%, 100 mL	Erlenmeyer flasks, 250-mL, 15
Iron(III) chloride solution, $FeCl_3$, 0.1 M, 50 mL	Filter paper, 15 sheets
Salicylic acid, $HO–C_6H_4–CO_2H$, 8 g	Funnels, 15
Sulfuric acid, concentrated, H_2SO_4, 2 mL	Graduated cylinders, 10-mL, 15
Distilled water and wash bottles, 15	Hot plates, 5–7*
Ice, crushed	Pasteur pipets or eyedroppers, 5
Melting point apparatus or Thiele-Dennis tube	Ring (support) stands and clamps, 15
Beakers, 50-mL, 15	Stirring rods, 15
Beakers, 250-mL, 30	Test tubes, 13 × 100 mm, 45
Beral-type pipets, graduated, 60	Thermometers, 15
Boiling stones	Watch glasses, 15

*Several groups may share hot plates.

Preparation of Solutions

Iron(III) Chloride, 0.1 M: Dissolve 2.0 grams of iron(III) chloride hexahydrate ($FeCl_3 \cdot 6H_2O$) in about 50 mL of distilled or deionized water. Stir to dissolve and dilute to 100 mL with water.

Safety Precautions

Concentrated sulfuric acid is severely corrosive to eyes, skin, and body tissue. Notify the teacher and clean up all spills immediately. Acetic anhydride is a corrosive liquid and the vapors are highly irritating. The liquid is flammable and a strong lachrymator—contact with the liquid will cause severe eye irritation. Work with acetic anhydride in the hood or in a well-ventilated lab only. Do not inhale the vapors. Salicylic acid is moderately toxic by ingestion. Avoid contact of all chemicals with eyes and skin. Wear chemical splash goggles, chemical-resistant gloves, and a chemical-resistant apron. Remind students that all chemicals prepared in the lab are for laboratory use only and should never be removed from the lab. The aspirin prepared in this lab may be impure and contaminated with chemicals that could be dangerous if ingested. Please review current Material Safety Data Sheets for additional safety, handling, and disposal information. Remind students to wash their hands thoroughly with soap and water before leaving the laboratory.

Disposal

Please consult your current *Flinn Scientific Catalog/Reference Manual* for general guidelines and specific procedures governing the disposal of laboratory waste. The dry solids may be disposed of in the trash according to Flinn Suggested Disposal Method #26a.

Teacher's Notes

Lab Hints

- For best results, schedule at least two 50-minute lab periods for this experiment. In addition, at least two hours (overnight is best) are needed to dry the aspirin. The mass of the aspirin and its melting point can be determined after school or during a free period the next day. The products may be stored in a lab drawer until the next regularly scheduled lab (i.e., the following week).

- Concentrated sulfuric acid and acetic anhydride are corrosive liquids. We recommend setting out and dispensing these chemicals in a central and *supervised* location. This is best done in the fume hood! Place the reagent bottles on a demonstration tray or a laboratory spill mat (LabMat™) to contain any chemical spills. To further reduce spillage, tape a small test tube to the side of each reagent bottle. Store a pipet in the test tube for students to use.

- The procedure calls for scraping the product from the filter paper onto a watch glass prior to drying the crystals. The filter paper tends to retain some acetic acid from the reaction mixture and does not dry out well, even overnight.

- Thiele-Dennis tubes are designed to give excellent convection and heat transfer for melting point and boiling point determinations. The unique design creates convection currents when the oil inside the tube is heated, allowing the oil to flow continuously through the tube without stirring or shaking. The recommended heating fluid is silicone oil or vegetable oil. *Fill the tube to the level shown—the oil will expand when heated.*

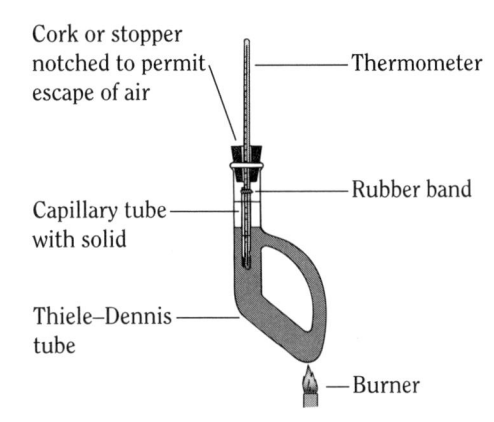

Figure 4. Design and Use of a Thiele-Dennis Tube.

- Melting points and boiling points of organic compounds may be found in the *Merck Index, CRC Handbook of Chemistry and Physics,* and *Lange's Handbook of Chemistry,* etc. Melting points may also be found online by looking up Material Safety Data Sheets. Section 9 of the MSDS lists common physical and chemical properties, including melting point, boiling point, density, etc. Visit the Flinn Web site www.flinnsci.com to download current MSDS for all Flinn chemicals.

- Compare the acidity of salicylic acid and acetylsalicylic acid (aspirin) by testing the pH of solutions (see step 19 in the *Procedure*) with narrow range (3.0–5.5) pH paper. Salicylic acid is more acidic than aspirin (pH 3.0 versus 3.5).

Never heat a closed system! Make sure the cork or stopper in the Thiele-Dennis tube is notched to allow air to escape.

Teacher's Notes

Teacher Notes

Teaching Tip

- See the *Supplementary Information* on pages 65–66 in *Acids and Bases,* Vol. 13 in the *Flinn ChemTopic™ Labs* series for a procedure to analyze the purity of aspirin by acid–base titration.

Answers to Pre-Lab Questions *(Student answers will vary.)*

1. Acetic anhydride is a *lachrymator*. What is a lachrymator and what safety precautions should be followed when working with acetic anhydride?

   A lachrymator is a substance that causes tearing and severe eye irritation. Avoid breathing the vapor and use the substance in the hood or in a well-ventilated lab only. Wear chemical splash goggles and chemical-resistant gloves and apron.

2. What is concentrated sulfuric acid used for in this experiment? What are the hazards of working with concentrated sulfuric acid?

   Concentrated sulfuric acid is used as an acid catalyst for the esterification reaction of salicylic acid with acetic anhydride. It is extremely corrosive and will cause severe burns. Be very careful when working with concentrated acids and avoid contact with eyes and skin. Notify the teacher immediately if any acid, even a few drops, is spilled. Wear chemical splash goggles and chemical-resistant gloves and apron.

3. Calculate (a) the molar mass of salicylic acid and acetic anhydride and (b) the number of moles of each that will be used in this experiment. *Note:* The density of acetic anhydride is 1.08 g/mL.

   (a) Salicylic acid (SA)
   $C_7H_6O_3$, *138 g/mole*

   Acetic anhydride (AA)
   $C_4H_6O_3$, *102 g/mole*

   (b) $\dfrac{0.50 \text{ g salicylic acid}}{138 \text{ g/mole}} = 0.0036 \text{ mole SA}$

   $\dfrac{1.0 \text{ mL acetic anhydride} \times 1.08 \text{ g/mL}}{102 \text{ g/mole}} = 0.011 \text{ mole AA}$

4. Define the term *limiting reactant*. Complete the following statement: The maximum number of moles of aspirin that can be obtained in this experiment is equal to the number of moles of **salicylic acid** used.

   *The limiting reactant in a chemical reaction is the reactant that is present in the smallest number of moles **based on the stoichiometric mole ratios** for the reactants. The number of moles of the limiting reactant determines (i.e., limits) the maximum amount of product that can be obtained.*

5. Determine the chemical formula of acetylsalicylic acid and calculate its molar mass.

   Acetylsalicylic acid, $C_9H_8O_4$, *180 g/mole*

Synthesis of Aspirin

Teacher's Notes

Sample Data

Student data will vary.

Data Table

Mass of salicylic acid used	0.51 g
Mass of watch glass	1.01 g
Mass of watch glass and acetylsalicylic acid	1.41 g
Melting point of acetylsalicylic acid	120–122 °C
Results of Fe^{3+} Tests	
Salicylic acid	Dark purple color
Reaction product	Light yellow—color of Fe^{3+} ions
Crushed aspirin	Slight pink (peach) coloration

Answers to Post-Lab Questions *(Student answers will vary.)*

1. Calculate the number of moles of salicylic acid used in this experiment.

   $$\frac{0.51 \text{ g salicylic acid}}{138 \text{ g/mole}} = 0.0037 \text{ mole}$$

2. Calculate the maximum amount of acetylsalicylic acid in grams that could be obtained from this amount of salicylic acid. This is the theoretical yield. *Hint:* See *Pre-Lab Questions* #4 and 5.

   $$0.0037 \text{ mole} \times 180 \text{ g/mole} = 0.67 \text{ g aspirin}$$

3. Determine the mass of aspirin obtained in this experiment and calculate the *percent yield*.

   $$\text{Percent yield} = \frac{\text{Actual yield}}{\text{Theoretical yield}} \times 100\%$$

   $$(0.40 \text{ g}/0.67 \text{ g}) \times 100 = 60\%$$

4. Iron(III) ions are used as a qualitative test for phenols (aromatic compounds containing an –OH functional group). (a) What compound was used as a *positive control* for the Fe^{3+} test in this experiment? (b) Did the reaction product give a positive or negative test result with Fe^{3+} ions? Explain.

   (a) Salicylic acid, which contains a phenolic –OH group in its structure, served as a control sample that would give a positive test. Salicylic acid gave a dark purple product when iron(III) chloride solution was added.

Teacher Notes

(b) The aspirin product gave a negative test with Fe^{3+} ions—no color change was observed. Aspirin does not contain an –OH functional group attached to the aromatic ring. The –OH group in salicylic acid was converted to an acetyl ($-OCOCH_3$) group.

5. Old aspirin tablets often have a faint vinegar smell (acetic acid) and give a positive test with iron(III) ions. Write a balanced chemical equation for the *hydrolysis* of aspirin (reaction of aspirin with water) to explain these observations.

   Hydrolysis of aspirin gives salicylic acid (positive Fe^{3+} test) and acetic acid (vinegar smell).

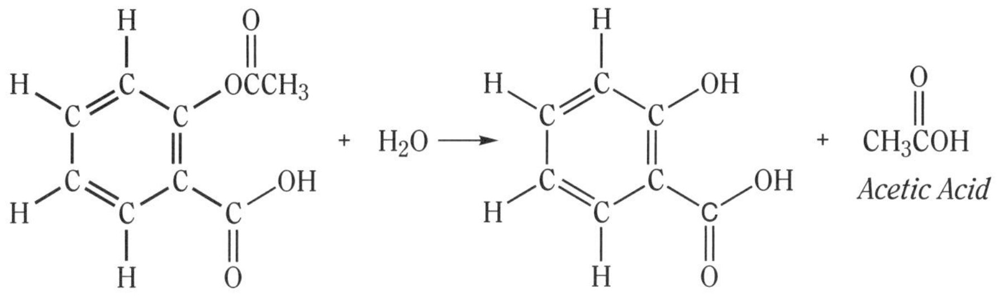

6. Acetic anhydride was used in excess in this experiment. What does this mean, and how was the excess acetic anhydride decomposed at the end of the reaction?

   The number of moles of acetic anhydride was greater than the number of moles that would react completely with the amount of salicylic acid in the reaction mixtures. Excess acetic anhydride was therefore left over in the reaction mixture after all the salicylic acid had been converted to product. The acetic anhydride remaining at the end of the reaction was decomposed by adding water to the reaction mixture prior to crystallization.

7. Look up the melting points of salicylic acid and aspirin (acetylsalicylic acid) in a reference book or online and compare with the melting point of the reaction product.

   Literature melting points:

   Salicylic acid, mp 157–159 °C

   Acetylsalicylic acid, mp 135 °C (dec)

   Melting point of product, 120–122 °C

   The product does not appear to be pure. The most likely contaminant is unreacted salicylic acid—the melting point of a mixture is lower than that of the pure substance.

Teacher's Notes

Teacher Notes

Page 1 – **Steam Distillation of Cinnamon**

Teacher Notes

Steam Distillation of Cinnamon
Cinnamaldehyde and Oil of Cinnamon

Introduction

Organic chemistry has its roots in the study of natural products. *Essential oils,* which have been used since ancient times as perfumes, flavorings and even medicines, are an interesting class of natural products. So-called because they seem to contain the odor and flavor "essence" of a plant, essential oils are volatile organic liquids obtained from flowers, leaves, twigs, fruits, nuts or seeds.

Concepts

- Essential oil
- Steam distillation
- Solvent extraction
- Aldehyde functional group

Background

There are three main methods for obtaining essential oils from plants: steam distillation, extraction with a solvent, and expression or cold pressing. *Steam distillation,* which was first used in the 13th century in France and Spain, involves heating the plant components with water or steam. The liquid that is collected when the resulting vapor condenses is called the distillate and consists of an immiscible mixture of water and oil—the essential oil. The essential oil can be separated from water by dissolving it in an organic solvent and then evaporating the solvent.

The boiling point of a pure substance is the temperature at which the vapor pressure of the liquid is equal to the atmospheric pressure (760 mm Hg for standard atmospheric pressure). When a mixture of two immiscible liquids, such as water and oil, is heated, *each liquid exerts its vapor pressure independently of the other liquid*. The total vapor pressure (P_{total}) at a given temperature is therefore equal to the *sum* of the vapor pressures of the two immiscible liquids. Equation 1 describes the total vapor pressure for a mixture of an essential oil and water.

$$P_{total} = P°_{H_2O} + P°_{Oil} \qquad \textit{Equation 1}$$

A mixture of two immiscible liquids will boil when their combined vapor pressure is equal to 760 mm Hg. The boiling point of a mixture of water and essential oil will thus be *less than* the boiling point of either component separately. This is a great advantage in the distillation of natural products that have high boiling points and would decompose at high temperatures. The boiling point of the distillate will remain constant during steam distillation as long as two phases are present. In order for steam distillation to be practical, an organic liquid must be volatile—it must have a vapor pressure of at least 5 mm Hg at 100 °C.

The amount of water and essential oil that will co-distill during steam distillation can be determined using Equation 1 and the ideal gas law. The *mole ratio* of oil and water in the distillate is equal to their relative vapor pressures ($P°$) at the boiling point (Equation 2, n = number of moles).

$$\frac{n(\text{oil})}{n(\text{H}_2\text{O})} = \frac{P°_{oil}}{P°_{H_2O}} \qquad \textit{Equation 2}$$

Steam Distillation of Cinnamon

Steam Distillation of Cinnamon – Page 2

Multiplying each term in Equation 2 by the molar mass (MM, g/mole) of the substance gives the mass of water in grams that will co-distill with an essential oil (Equation 3).

$$\frac{n(\text{oil}) \times \text{MM(oil)}}{n(\text{H}_2\text{O}) \times \text{MM (H}_2\text{O})} = \frac{\text{mass of oil (g)}}{\text{mass of water (g)}} = \frac{P°_{\text{oil}} \times \text{MM (oil)}}{P°_{\text{H}_2\text{O}} \times \text{MM (water)}} \quad \textit{Equation 3}$$

Cinnamon is obtained from the inner bark of *Cinnamomum zeylanicum*, a small evergreen tree that is native to Sri Lanka (formerly Ceylon) and India. Related species of *Cinnamomum* are found in China, Indonesia, and Vietnam. Cinnamon has been used as a spice for at least 6000 years and was highly prized in ancient India, China, Egypt, Rome, and Greece. Oil of cinnamon is obtained from cinnamon bark by steam distillation. By the early 1700s, there was a large trade in cinnamon oil throughout Europe and Southeast Asia (the East Indies). In 1834, the French chemists Jean-Baptiste Dumas and Eugène Péligot analyzed the composition of oil of cinnamon and identified cinnamaldehyde as the major component (70–80% of the oil). *Cinnamaldehyde,* which has the distinctive odor and taste of cinnamon, is classified as an aromatic aldehyde—it contains an *aldehyde* functional group and has a benzene (aromatic) ring (Figure 1). The compound has natural antimicrobial properties but is highly irritating to the skin. Minor components of oil of cinnamon include eugenol (10–20%) and cinnamic acid (5–10%). Like cinnamaldehyde, most components of essential oils are volatile organic compounds containing carbon, hydrogen, and oxygen.

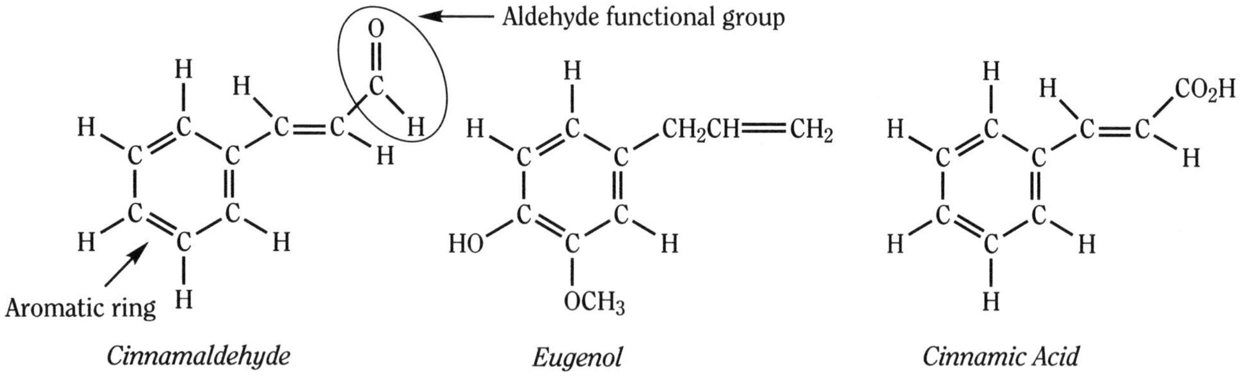

Figure 1. Components of Oil of Cinnamon.

Aromatic aldehydes are common ingredients in other essential oils and natural flavors. Examples include benzaldehyde (oil of bitter almond), cuminaldehyde (cumin seed oil), and vanillin (vanilla extract). See Figure 2.

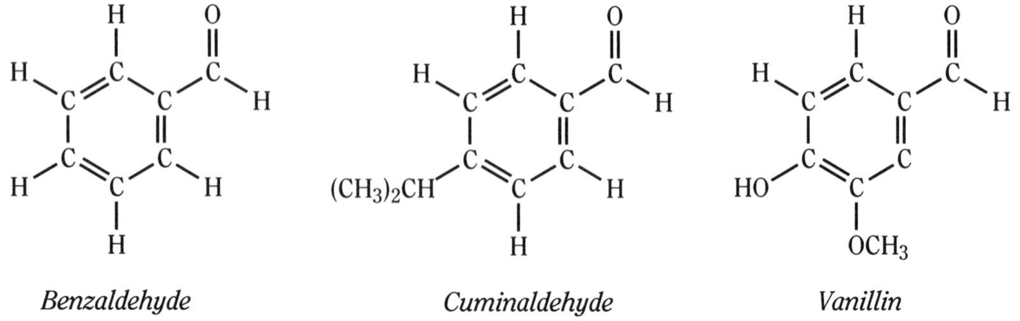

Figure 2. Aromatic Aldehydes in Nature.

Teacher Notes

Page 3 – Steam Distillation of Cinnamon

Experiment Overview

The purpose of this experiment is to isolate oil of cinnamon by steam distillation and to study its chemical properties. The mixture of oil and water in the distillate will be extracted with hexane to dissolve the oil and the solvent will then be evaporated. The presence of cinnamaldehyde, the main component in oil of cinnamon, will be identified using qualitative tests for organic functional groups.

Pre-Lab Questions

1. What is an essential oil?

2. (a) What are the specific requirements for a compound to steam distill with water?
   (b) What are the advantages of steam distillation for the isolation of natural products?

3. Limonene, the principal component of citrus peel oil, is a hydrocarbon ($C_{10}H_{16}$). At standard atmospheric pressure (760 mm Hg), a mixture of limonene and water steam distills at 98 °C. The vapor pressure of water at this temperature is 707 mm Hg. Calculate: (a) the vapor pressure of limonene at 98 °C; (b) the molar mass of limonene; and (c) the mass of water in grams that will co-distill with one gram of limonene during steam distillation.

Materials

Cinnamaldehyde, 0.5 mL	Erlenmeyer flask, 125-mL
Cinnamon sticks, fresh, 10 g	Heating mantle and variable transformer
Hexane, 15 mL	
Schiff reagent, 2 mL	Round-bottom flask, 250-mL
Sodium chloride, 10 g	Round-bottom flask, 100-mL
Water	Rubber tubing, 2 pieces
Custom distillation head (condenser and adapters)	Separatory funnel, 125-mL
Hot plate	Stirring rod
Thermometer and thermometer adapter	Stopcock grease (optional)
Beral pipets, 3	Support (ring) stands, 2
Beaker, 400-mL	Test tube, large
Clamps, 2	Test tubes, small, 2
Erlenmeyer flasks, 50-mL, 2	Weighing dish, large

Safety Precautions

Oil of cinnamon and cinnamaldehyde are severe skin irritants. Hexane is a flammable liquid and a dangerous fire risk—do not use around flames. Avoid contact of all chemicals with eyes and skin. Wear chemical splash goggles, chemical-resistant gloves, and a chemical-resistant apron. Wash hands thoroughly with soap and water before leaving the laboratory.

Steam Distillation of Cinnamon – Page 4

Procedure

1. Obtain about 10 g of fresh cinnamon sticks in a large weighing dish and break the sticks into small pieces.

2. Add the broken cinnamon pieces to a 250-mL round-bottom flask and fill the flask about one-half full with water. This is the distilling flask.

3. Place the distilling flask into a heating mantle and clamp the flask to a support (ring) stand.

4. Place the one-piece custom distillation head (or a three-way adapter, followed by a condenser and a second adapter) into the distilling flask. See Figure 3.

5. Place the thermometer holder into the adapter in the distillation head. Carefully insert the thermometer so that the thermometer bulb rests just below the side arm that leads to the condenser.

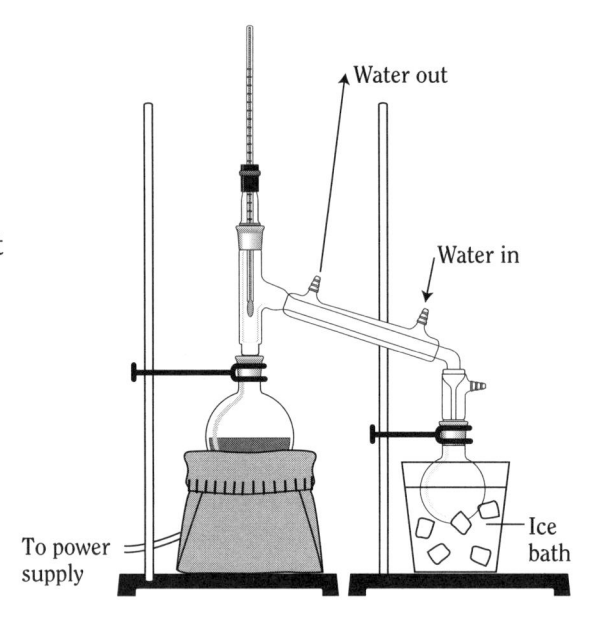

Figure 3. Custom Distillation Apparatus.

6. Place a 100-mL round-bottom flask into the side-arm adapter attached to the distillation head. Gently clamp the receiving flask to a support stand. Cool the flask in an ice bath.

7. Attach one piece of rubber tubing from the inlet nozzle on the condenser to the faucet, and place the second piece of rubber tubing from the outlet nozzle on the condenser into the sink to drain.

8. Turn on the faucet to allow water to flow through the condenser. Make sure the rubber tubing connections are tight and check for leaks.

9. Plug the heating mantle into the variable transformer and turn on the power to begin heating the flask.

10. When the water in the flask boils, the vapor will rise in the distillation apparatus and condense in the condenser. The liquid will collect in the receiving flask.

11. Continue distilling the mixture for 45 minutes, until about 30 mL of distillate has been collected. Record the boiling point of the distillate and its appearance in the data table.

12. Turn off the heating mantle. Allow the apparatus to cool and then remove the receiving flask.

13. Pour the distillate into a large test tube and add solid sodium chloride until the solution is saturated. Pour the liquid (*not* the solid) into a separatory funnel.

Teacher Notes

For best results, have students apply a small amount of stopcock grease to all glass joints before beginning the experiment. If a three-way adapter will be used with a separate condenser and outlet adapter, the condenser must also be clamped to a ring stand or securely fastened to the adapter with a rubber band.

Teacher Notes

14. Add about 7 mL of hexane to the separatory funnel and shake. Run off the lower aqueous layer back into the large test tube, and collect the upper hexane layer in a 125-mL Erlenmeyer flask.

15. Return the aqueous layer back into the separatory funnel, and repeat step 14. Combine the hexane extracts in the same Erlenmeyer flask.

16. Working in the hood, place a stirring rod in the hexane solution and heat the solution on a hot plate at a medium setting. Allow all of the hexane to evaporate.

17. *(Optional)* Allow the hexane to evaporate at room temperature overnight.

18. Cool the flask to room temperature. Observe the appearance of oil of cinnamon and carefully waft the vapor to your nose to detect the odor. In the data table, record the appearance and odor of the essential oil.

19. Obtain two small test tubes and place about 1 mL of Schiff reagent into each test tube.

20. Add one drop of pure cinnamaldehyde to the first test tube and one drop of the essential oil to the second test tube. Record observations in the data table.

Name: _____

Class/Lab Period: _____

Steam Distillation of Cinnamon

Data Table

Boiling Point and Appearance of Distillate	
Oil of Cinnamon—Appearance and Odor	
Results of Schiff Test for Aldehydes	
Cinnamaldehyde	
Oil of Cinnamon	

Post-Lab Questions *(Use a separate sheet of paper to answer the following questions.)*

1. Describe the appearance of the distillate. Is cinnamaldehyde (oil of cinnamon) less dense or more dense than water?

2. Compare the odor of cinnamon oil versus cinnamon sticks. What factors may account for the difference in smell?

3. Schiff's reagent is used as a qualitative test for aldehydes. What compound was used as a positive control or reference compound for this test? Describe the appearance of a positive test result. Did oil of cinnamon give a positive test?

4. (a) What was the boiling point of the distillate? (b) The vapor pressure of cinnamaldehyde at this temperature is about 5 mm Hg. Assuming that the barometric pressure was 760 mm Hg, use Equations 1 and 3 in the *Background* section to calculate the mass of cinnamon oil that would co-distill with 30 g of water during steam distillation.

5. Why is the boiling point of a mixture of an essential oil and water always less than 100 °C?

6. The word oil is a general term that is used to describe at least three different types of organic compounds: Vegetable oils (such as olive oil or canola oil), petroleum, and essential oils. Describe the *general* structure or composition of vegetable oils and petroleum.

7. Essential oils and other natural products are considered secondary metabolites—they are not responsible for the primary structure and function of a cell. What are some possible biochemical functions for an essential oil?

Questions 6 and 7 may be assigned for extra credit. This information is readily available via an online search.

Teacher's Notes
Steam Distillation of Cinnamon

Master Materials List *(for a class of 30 students working in pairs)*

Cinnamaldehyde, 10 mL
Cinnamon sticks, fresh, 150 g
Hexane, 250 mL
Schiff's reagent, 30 mL
Sodium chloride, 150 g
Water
Custom distillation heads (condenser and adapters), 15
Hot plates, 5–7*
Thermometers and thermometer adapters, 15
Beral pipets, 45
Beakers, 400-mL, 15
Clamps, 30
Erlenmeyer flasks, 50-mL, 30

Erlenmeyer flasks, 125-mL, 15
Heating mantles and variable transformers, 15
Round-bottom flasks, 250-mL, 15
Round-bottom flasks, 100-mL, 15
Rubber tubing, 30 pieces
Separatory funnels, 15
Stirring rods, 15
Stopcock grease (optional)
Support (ring) stands, 30
Test tubes, small, 30
Test tubes, large, 15
Weighing dishes, large, 15

*Several groups may share hot plates.

Safety Precautions

Oil of cinnamon and cinnamaldehyde are severe skin irritants. Hexane is a flammable liquid and a dangerous fire risk—do not use around flames. Avoid contact of all chemicals with eyes and skin. Wear chemical splash goggles, chemical-resistant gloves, and a chemical-resistant apron. Please review current Material Safety Data Sheets for additional safety, handling, and disposal information. Remind students to wash hands thoroughly with soap and water before leaving the laboratory.

Disposal

Please consult your current *Flinn Scientific Catalog/Reference Manual* for general guidelines and specific procedures governing the disposal of laboratory waste. The reaction product may be disposed of down the drain with cold running water according to Flinn Suggested Disposal Method #26b.

Lab Hints

- For best results, schedule *at least* two 50-minute lab periods for completion of this activity. The *Pre-Lab Questions* may be assigned as homework in preparation for lab and should be reviewed the day before lab. This lab is best suited for an advanced or honors chemistry class. Consider setting up one steam distillation apparatus to demonstrate the setup. Show students how to carefully adjust two clamps so that there is no tension on the necks of the flasks, and remind students that water flows in the bottom of the condenser and out the top. For best results, use stopcock grease on all glass joints.

- There are two good stopping points if the lab cannot be finished in the scheduled lab time. After extracting the oil from the distillate, save the hexane extract before evaporating the solvent. Testing the oil of cinnamon will only take a few minutes. This step can easily be completed on a second day of lab.

Teacher's Notes

- Demonstrate the proper use of a separatory funnel before beginning the experiment. Remind students that they must always invert the stoppered separatory funnel and vent the vapor *before* shaking the funnel, and then again after every shake. Failure to vent the funnel may result in pressure building up and solvent spraying out of the separatory funnel when the stopper is removed.

- For best results, always use fresh cinnamon sticks for steam distillation. Cinnamaldehyde and other components of cinnamon oil are volatile liquids and may evaporate from opened packages of cinnamon.

- If purchasing a classroom set of glassware for organic chemistry labs is not realistic, schedule this lab concurrently with another experiment. Divide up the class so that half the class does this experiment one week, the other half the following week.

- Cinnamon sticks are the dried inner bark of the cinnamon tree. The composition of an essential oil varies for different parts of the plant. The main component of cinnamon leaf oil, for example, is eugenol rather than cinnamaldehyde, and the oil is more harsh (less subtle).

- Schiff's reagent is used as a qualitative test for aldehydes. It consists of an indicator dye, fuchsin hydrochloride, in a saturated solution of sulfur dioxide. Sulfur dioxide decolorizes the dye. When an aldehyde is added to Schiff's reagent, it reacts with the SO_2 and thus restores the deep reddish purple color of the dye.

- Limonene, the main component of citrus peel oil, may be obtained by steam distillation of orange peel.

Teaching Tips

- Essential oils are considered secondary metabolites. Unlike proteins, carbohydrates, and lipids, essential oils are not involved in the primary structure, function or metabolism of plants. The biological function of essential oils is not well understood. In general, essential oils seem to help plants adapt to their environments. Many leaf oils and root oils are natural pesticides, protecting plants against insects and parasites. Essential oils from flowers may help plants attract insects for pollination. Many essential oils also have natural antifungal and antibacterial ability.

- Cinnamaldehyde is an intermediate in the "shikimic acid pathway" for the synthesis of natural products (Figure 4). The immediate biological precursor of cinnamaldehyde is cinnamic acid, which is obtained from the amino acid phenylalanine by loss of ammonia (NH_3). Cinnamic acid is also the precursor of many phenolics, such as eugenol.

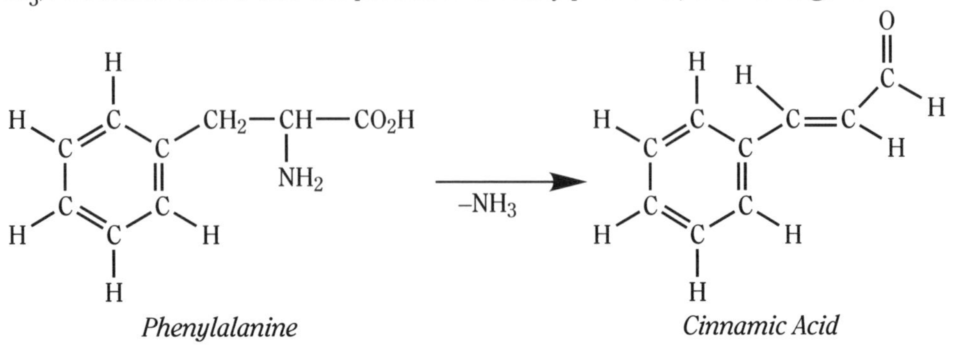

Phenylalanine → (–NH_3) → *Cinnamic Acid*

Figure 4. Biosynthesis of Cinnamaldehyde.

Teacher's Notes

Answers to Pre-Lab Questions *(Student answers will vary.)*

1. What is an essential oil?

   Essential oils are volatile organic liquids obtained from plants. They are natural "flavor and fragrance" chemicals in plants.

2. (a) What are the specific requirements for a compound to steam distill with water?

   (b) What are the advantages of steam distillation for the isolation of natural products?

   (a) In order for a compound to steam distill with water it must be insoluble in water (immiscible) and volatile, with a vapor pressure of at least 5 mm Hg at 100 °C.

   (b) Steam distillation occurs at temperatures less than 100 °C—this is very useful for natural products that would decompose if heated to their normal boiling points (≥ 200 °C).

3. Limonene, the principal component of citrus peel oil, is a hydrocarbon ($C_{10}H_{16}$). At standard atmospheric pressure (760 mm Hg), a mixture of limonene and water steam distills at 98 °C. The vapor pressure of water at this temperature is 707 mm Hg. Calculate: (a) the vapor pressure of limonene at 98 °C; (b) the molar mass of limonene; and (c) the mass of water in grams that will co-distill with one gram of limonene during steam distillation.

   (a) $P_{total} = P°_{H_2O} + P°_{Oil}$

   Vapor pressure of limonene = 760 − 707 mm Hg = 53 mm Hg at 98 °C

   (b) Molar mass of limonene ($C_{10}H_{16}$) = 136 g/mole

   Molar mass of water (H_2O) = 18 g/mole

   (c) $\dfrac{\text{mass of limonene (g)}}{\text{mass of water (g)}} = \dfrac{P°_{oil} \times MM\ (oil)}{P°_{H_2O} \times MM\ (water)}$

   $\dfrac{1\ g\ \text{limonene}}{x\ g\ \text{water}} = \dfrac{(53\ mm\ Hg)(136\ g/mole)}{(707\ mm\ Hg)(18\ g/mole)}$

   $x = 1.8\ g\ \text{water per 1 g of limonene}$

Steam Distillation of Cinnamon

Teacher's Notes

Sample Data

Student data will vary.

Data Table

Boiling Point and Appearance of Distillate	Boiling point = 99–100 °C. The distillate looked oily and separated into two layers in receiving flask.
Oil of Cinnamon—Appearance and Odor	The oil was a pale yellow liquid with a strong cinnamon odor.
Results of Schiff Test for Aldehydes	
Cinnamaldehyde	The sample gave a deep magenta (purple) solution after 2 min.
Oil of Cinnamon	Same as above—a deep purple product.

Answers to Post-Lab Questions *(Student answers will vary.)*

1. Describe the appearance of the distillate. Is cinnamaldehyde (oil of cinnamon) less dense or more dense than water?

    The distillate looked like oily water. It was two layers. When the liquid in the condenser collected in the receiving flask, the oil seemed to be the bottom layer. Oil of cinnamon is thus more dense than water. **Note to teachers:** *Most essential oils are less dense than water. Oil of cinnamon is an exception.*

2. Compare the odor of cinnamon oil versus cinnamon sticks. What factors may account for the difference in smell?

    Cinnamon oil has a stronger smell than cinnamon sticks. That is because there may be many more components of the odor of fresh natural cinnamon in addition to the essential oil. The other components may modify or balance the odor of the essential oil.

3. Schiff's reagent is used as a qualitative test for aldehydes. What compound was used as a positive control or reference compound for this test? Describe the appearance of a positive test result. Did oil of cinnamon give a positive test?

    Cinnamaldehyde is a known aldehyde. It served as a positive control for the test with Schiff's reagent. A positive test was marked by the formation of a dark magenta (reddish purple) solution. Oil of cinnamon gave a positive test.

Teacher's Notes

Teacher Notes

4. (a) What was the boiling point of the distillate? (b) The vapor pressure of cinnamaldehyde at this temperature is about 5 mm Hg. Assuming that the barometric pressure was 760 mm Hg, use Equations 1 and 3 in the *Background* section to calculate the mass of cinnamon oil that would co-distill with 30 g of water during steam distillation.

   (a) The boiling point was 99–100 °C.

   (b) Vapor pressure of water = 760 – 5 mm Hg = 755 mm Hg.

   (c) Cinnamaldehyde, C_9H_8O, molar mass 131 g/mole

$$\frac{\text{mass of cinnamaldehyde (g)}}{\text{mass of water (g)}} = \frac{P^o_{oil} \times MM\,(oil)}{P^o_{H_2O} \times MM\,(water)}$$

$$\frac{x \text{ grams cinnamaldehyde}}{30 \text{ g water}} = \frac{(5 \text{ mm Hg})(131 \text{ g/mole})}{(755 \text{ mm Hg})(18 \text{ g/mole})}$$

   x = 1.5 g cinnamaldehyde per 30 g of water

5. Why is the boiling point of a mixture of an essential oil and water always less than 100 °C?

   The combined vapor pressure of a mixture of two immiscible liquids is equal to the sum of their individual vapor pressures. Because a liquid will boil when the total vapor pressure is equal to atmospheric pressure, the boiling point of the mixture will be less than the boiling point of either component.

6. The word oil is a general term that is used to describe at least three different types of organic compounds: Vegetable oils (such as olive oil or canola oil), petroleum, and essential oils. Describe the *general* structure or composition of vegetable oils and petroleum.

   Vegetable oils such as olive oil or canola oil are triglycerides—they are esters formed by combination of one glycerol molecule with three fatty acids. Petroleum is a mixture of hydrocarbons (compounds containing only carbon and hydrogen).

7. Essential oils and other natural products are considered secondary metabolites—they are not responsible for the primary structure and function of a cell. What are some possible biochemical functions for an essential oil?

   Essential oils are responsible for the strong odors or fragrance of plants. Odors help plants attract insects for pollination. Also, some odors may actually repel insects that may be harmful. **Note to teachers:** *A good example is citronellol, which is the essential oil from lemongrass. Citronella candles are used to keep mosquitos away.*

Teacher's Notes

Teacher Notes

Teacher Notes

Cleaning with Charcoal
Turning Grape Juice into Water!

Introduction

Activated charcoal is widely used in water purification—whether in municipal water treatment plants or in aquariums—to remove impurities and contaminants. Demonstrate the "cleaning action" of charcoal by decolorizing grape juice!

Concepts

- Activated charcoal
- Adsorption
- Water treatment

Materials

Activated charcoal, 3 g

Grape juice, purple, 90 mL

Beaker, 250-mL

Filter paper (to fit funnel)

Funnel

Ring clamp and support stand

Stirring rod

Water goblet or Erlenmeyer flask, 125-mL

Weighing dish

Safety Precautions

Charcoal is a flammable solid. Wear chemical splash goggles whenever working with chemicals, heat or glassware in the laboratory. Please review current Material Safety Data Sheets for additional safety, handling, and disposal information.

Procedure

1. Pour 90 mL of purple grape juice into a 250-mL beaker.
2. Weigh about 3 g of activated charcoal into the weighing dish.
3. Add the charcoal to the grape juice and mix gently with a stirring rod.
4. Set up a glass funnel with a ring clamp on a support stand.
5. Fold qualitative filter paper into a cone and place in the funnel.
6. Set a water goblet or a 125-mL Erlenmeyer flask under the funnel.
7. Carefully pour the grape juice–charcoal mixture into the funnel. Collect the filtrate in the water goblet.
8. The filtrate should be a clear and colorless liquid—all traces of color and odor are removed by the charcoal.

Disposal

Please consult your current *Flinn Scientific Catalog/Reference Manual* for general guidelines and specific procedures governing the disposal of laboratory waste. The decolorized grape juice may be poured down the drain. Allow the charcoal filter to dry out, then dispose of in the trash according to Flinn Suggested Disposal Method #26a.

Demonstrations

Tips

- Transferring the grape juice from a wine glass into a water goblet (wine-to-water) may be more dramatic, but it is difficult to pour without dripping.

- In municipal water treatment plants, water is usually passed over a bed of activated carbon to remove contaminants. How much color can be removed by passing the grape juice once through a filter containing activated charcoal?

- Experiment with removing different kinds of colored impurities from water—try red food coloring, copper(II) sulfate, iron salts, etc.

Discussion

Activated charcoal (also called activated carbon) is obtained by the destructive (dry) distillation of wood or other plant and animal sources. The carbon residue obtained in this manner is "activated" by heating it with steam, oxygen or carbon dioxide. This process results in a finely divided solid with an extremely large surface area. The structure of activated charcoal is very porous (honeycomb-like) and thus has a high affinity for many substances, especially organic compounds, chlorine, and many gases.

Activated charcoal has been used since ancient times to remove undesirable contaminants from drinking water. Activated charcoal is an excellent adsorbent—a substance that is capable of attracting and binding the components of a mixture. (Adsorption refers specifically to the adhesion of atoms, ions or molecules onto the surface of another substance, usually a solid.) Because of its high adsorption capacity, activated charcoal is a critical component in all modern water and air purification systems. It is used to decolorize, deodorize, and clarify water. In this demonstration, activated charcoal removes natural organic indicator dyes that give grape juice its distinctive purple color and flavor.

The ability of activated charcoal to filter and remove contaminants from water depends on the particle size, surface area, and pore structure of the charcoal, as well as on pH, temperature, and the types of impurities.

Teacher Notes

Demonstrations

Teacher Notes

The Carbon Soufflé
Removing Water from Sugar!

Introduction

What is a carbohydrate? Carbon—plus water! Remove the water and whip up a carbon "soufflé" by reacting sugar with sulfuric acid. The mixture will turn brown and then black before expanding out the top of the beaker and solidifying. It even smells like cooking as the exothermic reaction fills the room with the odor of caramelizing sugar.

Concepts

- Carbohydrates
- Dehydration reaction
- Exothermic reaction

Materials

Sucrose (table sugar), $C_{12}H_{22}O_{11}$, 60 g
Sulfuric acid, concentrated, H_2SO_4, 18 M, 60 mL
Sodium carbonate, Na_2CO_3, 25 g
Beaker, Pyrex®, 250-mL
Graduated cylinder, 100-mL

Tongs
Balance
Demonstration tray (optional)
Paper towels
Stirring rod, glass

Safety Precautions

Concentrated sulfuric acid is severely corrosive to eyes, skin, and other tissue. Mixing with water may cause spraying and splattering and will release considerable heat. Use tongs to handle the product of the reaction. The carbon product will contain unreacted sulfuric acid. Neutralize acid spills and the carbon product with sodium carbonate. The steam produced by the reaction can cause burns. Do not stand over the reaction vessel or inhale the steam produced. The reaction will get extremely hot; allow the beaker to cool before handling. Perform this demonstration in a fume hood or a well-ventilated room. Wear chemical splash goggles, chemical-resistant apron, and chemical-resistant gloves. Please review current Material Safety Data Sheets for additional safety, handling, and disposal information.

Procedure

Please review all safety precautions. Perform this demonstration in a fume hood or in a well-ventilated lab only.

1. Add 60 grams of sucrose (table sugar) to a 250-mL Pyrex® beaker.

2. Place the beaker on a layer of paper towels (on a demonstration tray, if possible).

3. Using a 100-mL graduated cylinder, carefully measure 60 mL of concentrated sulfuric acid. (Any acid spills should be neutralized immediately with sodium carbonate.)

4. Slowly pour the sulfuric acid into the beaker containing sucrose.

5. Stir briefly with a glass stirring rod. Leave the stirring rod inside the beaker—it will help support the column of carbon produced.

"The Carbon Soufflé" is available as a chemical demonstration kit from Flinn Scientific, Catalog No. AP4422.

Demonstrations

6. Stand back and observe. In a few minutes the solution will start to bubble and expand. Steam will be visible coming out of the mouth of the beaker. The beaker will get very hot! The carbon "soufflé" reaction will take about 15 minutes from the time the sulfuric acid is added to when the carbon product has hardened.

7. Allow the beaker to cool. Follow the disposal procedure to neutralize the reaction mixture with sodium carbonate before removing the carbon product.

Disposal

Please consult your current *Flinn Scientific Catalog/Reference Manual* for general guidelines and specific procedures governing the disposal of laboratory waste. When the reaction is complete and the beaker is cool, sprinkle the carbon product with sodium carbonate to neutralize any remaining acid. Remove the carbon product from the beaker using tongs and rinse thoroughly under running water. Place the carbon product inside a plastic bag. Seal the bag and dispose of in the trash according to Flinn Suggested Disposal Method #26a.

Tip

- Use the time between adding the sulfuric acid and the start of the reaction to discuss the properties of sulfuric acid and carbohydrates.

Discussion

Plants combine carbon dioxide and water in the presence of chlorophyll and sunlight to produce glucose (sugar) and oxygen. The glucose is stored in the form of starch or other polysaccharides. The general formula for a sugar is $(C \cdot H_2O)_n$ for a monosaccharide, such as glucose or fructose ($C_6H_{12}O_6$), and $C_n \cdot (H_2O)_{n-1}$ for a disaccharide, such as sucrose ($C_{12}H_{22}O_{11}$). Carbohydrates may be thought of as "stored energy"—the energy is released when the carbohydrates are metabolized.

This demonstration provides a dramatic example of the amount of energy stored in a carbohydrate. Concentrated sulfuric acid is a strong dehydrating agent and will literally extract the water from the sugar and leave only carbon (Equation 1). Heat is generated during this dehydration step (–918.9 kJ/mol), and also from dilution of sulfuric acid with water (–40.6 kJ/mol, Equation 2). Some of the water is converted into steam because of the heat produced during the reaction.

$$C_{12}H_{22}O_{11}(s) \rightarrow 12C(s) + 11H_2O(l) \qquad \Delta H = -918.9 \text{ kJ/mol} \qquad \textit{Equation 1}$$

$$H_2SO_4(l) \rightarrow H_2SO_4 \cdot nH_2O(aq) \qquad \Delta H = -40.6 \text{ kJ/mol} \qquad \textit{Equation 2}$$

Teacher Notes

Sucrose is a nonreducing sugar. It is a disaccharide composed of one glucose unit connected to a fructose unit.

Teacher Notes

Feeling Blue
Organic Redox Reaction

Introduction

"Banish the blues" from your chemistry classroom! The reaction of a reducing sugar with an organic redox indicator shows us how the blues can come and go when things get a little shaken up.

Concepts

- Oxidation–reduction
- Reducing sugar
- Redox indicator

Materials

Potassium hydroxide solution, KOH, 1 M, 100 mL

Dextrose (glucose) solution, $C_6H_{12}O_6$, 0.4 M, 100 mL

Methylene blue solution, 1%, 1 mL

Beral-type pipet

Distilled water and wash bottle

Erlenmeyer flask, 500-mL with cap or stopper

Graduated cylinder, 100-mL

Safety Precautions

Potassium hydroxide solution is a corrosive liquid. Contact with skin may cause blisters. Potassium hydroxide is especially dangerous to the eyes and may be harmful if swallowed. Avoid contact of all chemicals with eyes and skin. Wear chemical splash goggles, chemical-resistant gloves, and a chemical-resistant apron. Please review current Material Safety Data Sheets for additional safety, handling, and disposal information.

Procedure

1. Using a 100-mL graduated cylinder, measure out 100 mL of 1 M potassium hydroxide solution and transfer to a 500-mL Erlenmeyer flask.

2. Rinse the graduated cylinder, then use it to measure out 100 mL of 0.4 M dextrose solution. Transfer the dextrose solution to the Erlenmeyer flask and swirl the flask to mix the solutions.

3. Add 3–4 drops of 1% methylene blue solution to the Erlenmeyer flask. Stopper the flask and swirl to mix the contents.

4. Allow the solution to stand undisturbed until it turns colorless—this may take a few minutes.

5. Shake the flask gently once or twice to restore the blue color, then allow it to sit undisturbed again until the blue color disappears.

6. Steps 4 and 5 may be repeated many times. Each time, shaking the flask causes the blue color to appear. Letting the flask sit undisturbed dissipates the color.

"Feeling Blue" is available as a chemical demonstration kit from Flinn Scientific (Catalog No. AP8653).

Demonstrations

Disposal

Please consult your current *Flinn Scientific Catalog/Reference Manual* for general guidelines and specific procedures governing the disposal of laboratory waste. Neutralize the resulting solution according to Flinn Suggested Disposal Method #10.

Tip

- Use the "Feeling Blue" demonstration to develop your students' observation and reasoning skills. What causes the blue color to reappear when the flask is shaken? How long does it take for the color to disappear after shaking? What factors influence how fast the color dissipates? Let students observe the blue "boundary layer" at the interface between the air in the flask and the solution below. The length of time the solution stays blue is directly proportional to the amount of shaking.

Discussion

Methylene blue (abbreviated MB) is an organic redox indicator that exists in two forms, a reduced form and an oxidized form. The reduced form of methylene blue (MB_{rd}) is colorless, while the oxidized form (MB_{ox}) is blue. The reduced form is easily converted to the oxidized form by mixing it with oxygen in the air (Equation 1). The oxidized form, in turn, can be converted back to the reduced form by treatment with a reducing agent.

$$MB_{rd} + O_2 \rightarrow MB_{ox} \qquad \text{Equation 1}$$
$$\text{Colorless} \qquad\qquad \text{Blue}$$

In this demonstration, the blue oxidized form (MB_{ox}) reacts with dextrose and potassium hydroxide to give the colorless, reduced form MB_{rd} (Equation 2).

$$MB_{ox} + \text{Dextrose} + KOH \rightarrow MB_{rd} \qquad \text{Equation 2}$$
$$\text{Blue} \qquad\qquad\qquad\qquad \text{Colorless}$$

Shaking the flask causes more oxygen to dissolve in the colorless solution, and methylene blue is re-oxidized according to Equation 1. The observation of the blue color at the interface between the air (oxygen) in the flask and the solution demonstrates that air (oxygen) enters the solution and causes the blue color.

The reaction pathway involves a series of fast reactions: potassium hydroxide converts dextrose (glucose) to a glucoside anion, oxygen in the flask goes into solution, and dissolved oxygen oxidizes the reduced (colorless) form of methylene blue to its oxidized blue form. The slow step is the reduction of MB_{ox} by the glucoside anion. The reaction pathway is summarized below:

$$\text{Glucose} + OH^- \xrightleftharpoons{\text{fast}} \text{Glucoside anion} + H_2O \qquad \text{Equation 3}$$

$$O_2(g) \xrightleftharpoons{\text{fast}} O_2(aq) \qquad \text{Equation 4}$$
$$\text{Dissolved oxygen}$$

$$MB_{ox} + \text{Glucoside anion} \xrightarrow{\text{slow}} MB_{rd} + \text{Glucose oxidation product(s)} \qquad \text{Equation 5}$$

Teacher Notes

Dextrose is D-glucose, a reducing sugar. This means that dextrose will reduce weak oxidizing agents, such as Cu^{2+} ions (Benedict's Test), Ag^+ ions (Tollens' Test), or an organic redox indicator (methylene blue in this demonstration).

Demonstrations

Teacher Notes

Kaleidoscope . . . Optical Activity
Rotation of Plane–Polarized Light

Introduction

A radially polarized filter is placed on an overhead projector stage and a regular (parallel) polarized filter is positioned above it. The projected image shows four quadrants, consisting of alternating light-dark-light-dark series of wedges. When an optically active solution (corn syrup) is poured between the filters, the image rotates and separates into a beautiful kaleidoscope of spectrum colors.

Concepts

- Chiral compounds
- Enantiomers
- Plane polarized light
- Optical activity

Materials

Polarizing film or filters, 6" × 6", 3 sheets
Cutting patterns to make "radially polarized filters"
Beaker, tall-form, 600-mL
Corn syrup, 400 mL
Overhead projector
Scissors
Transparent tape

Preparation

1. Tape one cutting pattern to opposite corners of one 6" × 6" piece of polarizing film. Using the cutting pattern as a template, cut 12 sharp wedges from the polarizing film. See Figure 1.

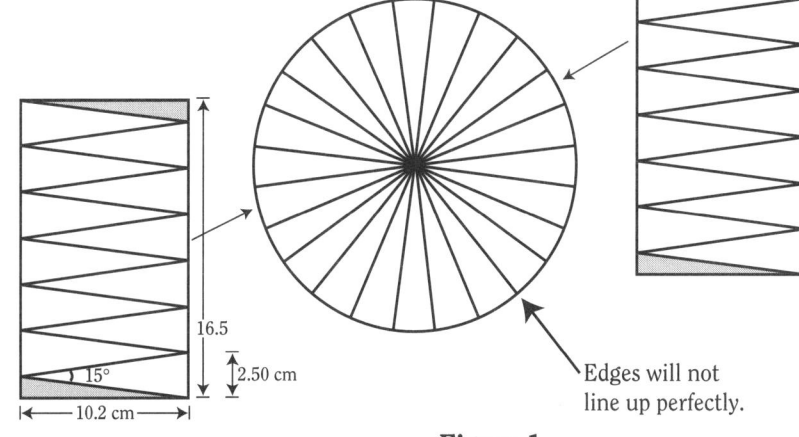

   Figure 1.
   (16.5, 15°, 2.50 cm, 10.2 cm, Edges will not line up perfectly.)

2. Repeat step 1 with the second cutting pattern and a second square of polarizing film. The plane of polarization in the second polarizing film *must be in the same direction* as that in the first filter when the wedges are cut. See the *Tips* section.

3. Rearrange the wedges into a pie shape and tape the wedges together using transparent tape. It's easiest to first place one wedge each at 0°, 90°, 180°, and 270° angles, and then to fill in the spaces between these four wedges. There will be a small space between adjacent wedges—this is not a problem.

This demonstration is available as a chemical demonstration kit from Flinn Scientific, Catalog No. AP8781.

Demonstrations

Procedure

1. Turn on the overhead projector light and place the radially polarized filter directly on the stage. Point out the spoke-like configuration of the filter wedges.

2. Clamp the plane polarized filter to the overhead projector or a ring stand so that the filter is about 20–30 cm above the radial filter pattern. Alternatively, hang the filter over the lens that directs the light onto the screen (Figure 2).

3. A bright "bow-tie" image should project onto the screen when the second (parallel) polarizer is placed over the radial filter pattern on the overhead projector stage.

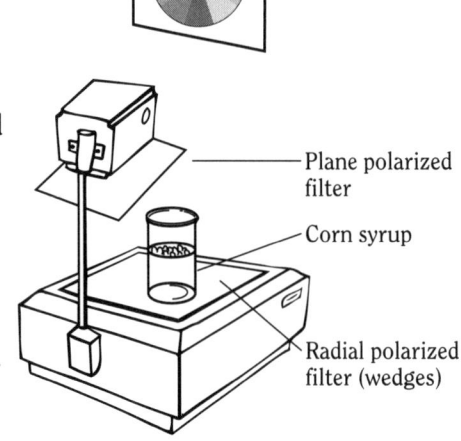

Figure 2.

4. Center a 600-mL tall-form beaker on top of the radial filter pattern, and gradually pour corn syrup into the beaker. Observe the spectrum and pattern of colors created on the projected image.

5. The dark and light wedges in the projected bow-tie image appear to rotate as the corn syrup is added. Each wedge's image becomes progressively brighter and then darker in sequence around the circular pattern as the plane polarized light passing through the wedge is first brought into, and then past, parallel alignment with the second polarizing filter above it.

6. Different colors of light become visible and progressively more distinct as more corn syrup is added. Blues and browns appear first, then purple, orange, yellow and green.

7. When the corn syrup depth is about 6–8 cm, the entire spectrum of colors will be visible two-fold around the circular image, producing a beautiful kaleidoscopic display. Steps 4–6 may be repeated many times.

Disposal

Please consult your current *Flinn Scientific Catalog/Reference Manual* for general guidelines and specific procedures governing the disposal of laboratory waste. The corn syrup may be stored for future use or flushed down the sink with plenty of excess water according to Flinn Suggested Disposal Method #26b. Save the polarizing filters for future use.

Tips

- Check that the polarizing filters are parallel before cutting the radial patterns. Place one 6" × 6" polarizing filter on the overhead projector stage, and lay the second filter on top of the first. If no light passes through (the top filter is dark), the filters are perpendicular—rotate the second filter 90° until light passes through. The filters will then be parallel.

- Cut a large hole in a piece of cardboard and place the cardboard on the overhead projector stage to block out peripheral light.

- Start the demonstration by placing a red filter on the overhead stage to eliminate different colors of light. This will help focus attention on the rotation of the light-dark-light-dark series of wedges in the bow-tie image.

Teacher Notes

Demonstrations

Teacher Notes

Discussion

Corn syrup is a concentrated solution of dextrose (D-glucose). *Dextrose* is an example of a *chiral* molecule, meaning it is not superimposable on its mirror image. The word chiral comes from the Greek word *chiros* (hand). In the same way that our left hands are not superimposable on our right hands (Figure 3), the two mirror images of a chiral molecule represent a pair of isomers called enantiomers. The two enantiomers of a chiral molecule cannot be interconverted by simply turning or rotating the molecules. Any organic compound containing a carbon atom attached to four different groups is a chiral molecule. Enantiomers have identical physical properties (melting point, boiling point, solubility, etc.). They differ, however, in their interaction with plane polarized light and their ability to rotate the plane of polarized light. Enantiomers may also have different physiological properties because of their interaction with chiral receptor molecules in the body. Like almost all biological molecules, including all carbohydrates and amino acids, glucose occurs in nature as a single enantiomer, not as a mixture of two enantiomers.

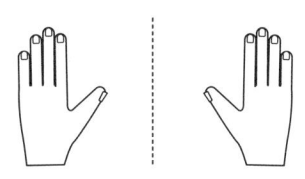

Figure 3.

Light is a form of electromagnetic radiation, consisting of oscillating electric and magnetic fields at right angles to each other. In normal light, the electric field vectors are oriented in all possible planes. Passing normal light through a "polarizer" converts light to *plane polarized light,* in which all of the electric field vectors lie in the same plane (Figure 4).

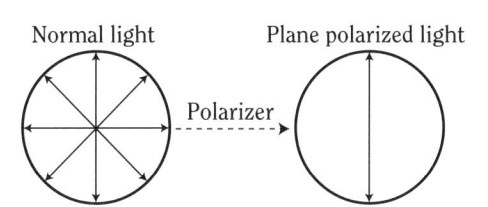

Figure 4. Plane Polarized Light

All molecules will rotate the plane of polarized light due to interaction of electrons with the electric field. For molecules that are not chiral, these rotations occur in all directions and thus cancel out. There is no net or overall rotation when plane polarized light passes through a solution of an achiral molecule. A pure enantiomer of a chiral compound, however, will rotate the plane of polarized light. The ability of a sample to rotate the plane of polarized light is called optical activity. Enantiomers have equal but opposite rotations. Thus, if one enantiomer rotates the plane of polarized light +25° (clockwise), the other enantiomer will rotate the plane –25° (counter-clockwise). Optical rotation depends on temperature, the concentration of solute, the path length, and the wavelength of light that is used.

The spectrum of colors in the rotating bow-tie image in this demonstration is produced because each wavelength of light is rotated by a different amount. The light that passes through the first radial polarizing filter is plane polarized, but it still contains all of the wavelengths of visible light. When the plane polarized light passes through the optically active corn syrup, each wavelength of light is rotated to a different degree. The plane polarized light, now separated into its various wavelengths, then passes through the second polarizing filter. Only one plane (and therefore one wavelength) of polarized light from each wedge will pass through this filter. Because the various wavelengths of light that pass through the polarizing filter wedges are bent to different plane angles, just one wavelength of light is able to pass through the second filter, and a rainbow of colors appears.

Examples of polarizing materials include the mineral calcite, Nicol prisms, and Polaroid films.

Kaleidoscope . . . Optical Activity

Demonstrations

Cutting Patterns for Radially Polarized Filter

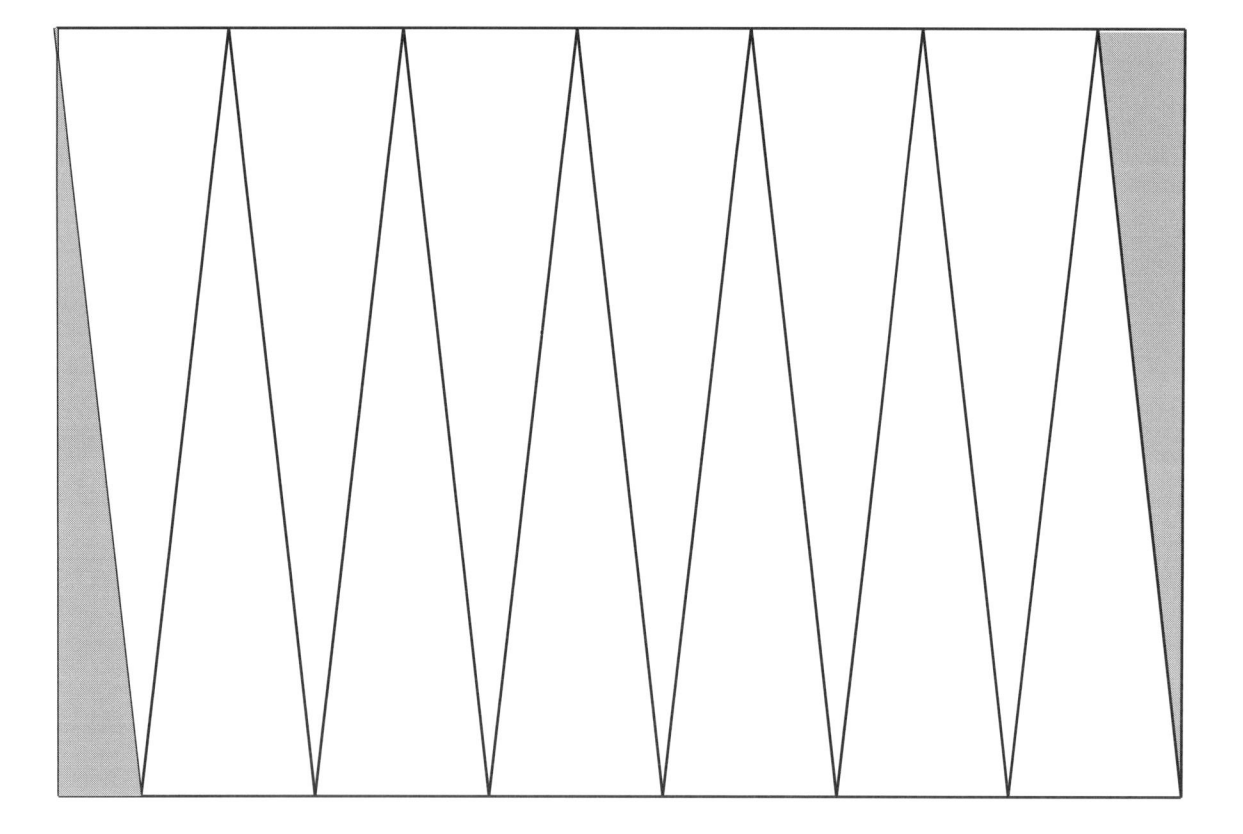

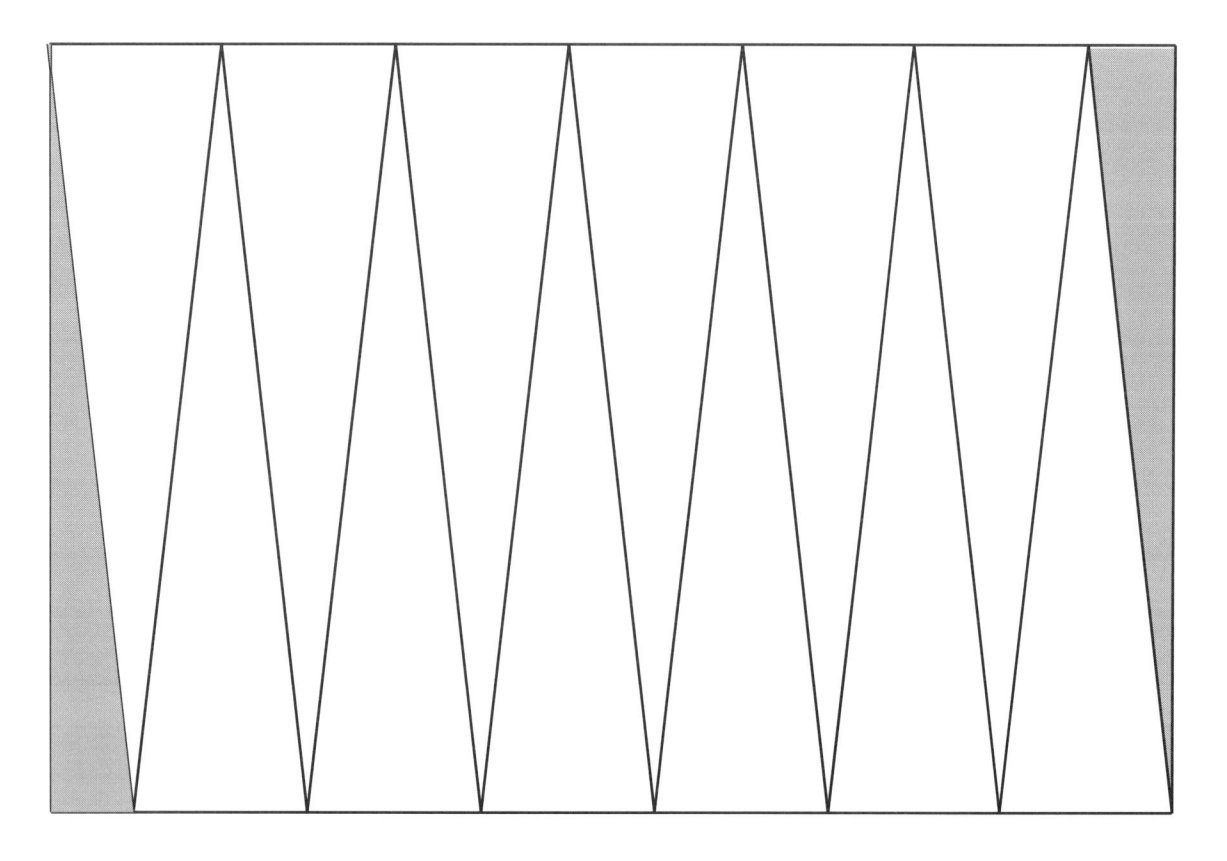

Teacher Notes

Demonstrations

Teacher Notes

Salt-Out the Red, White and Blue
Making a Three-Layer Liquid

Introduction

Create a beautiful three-layered liquid to demonstrate the salting-out effect and the miscibility of organic solvents with water.

Concepts

- Miscible and immiscible liquids
- Density
- Salting-out effect

Materials

Copper(II) sulfate, $CuSO_4 \cdot 5H_2O$, 0.5 g
Distilled or deionized water, 80 mL
Methyl alcohol, CH_3OH, 80 mL
Potassium carbonate, K_2CO_3, 55 g
Sudan III, 0.5 g
Toluene, $C_6H_5CH_3$, 80 mL
Balance

Beakers, 250-mL (1) and 150-mL, 3
Gas-collecting bottle, 240 mL
Microspatulas, 2
Pipet, 25- or 50-mL
Stirring rod
Weighing dish or cup, large

Safety Precautions

Methyl alcohol is a flammable solvent. It is toxic by ingestion and may cause blindness. Keep away from flames, heat, and all sources of ignition. Toluene is a combustible liquid and is moderately toxic by ingestion. Potassium carbonate is a corrosive solid. Perform this demonstration in a hood or in a well-ventilated lab only. Avoid contact of all chemicals with eyes and skin. Wear chemical splash goggles, chemical-resistant apron, and chemical-resistant gloves. Please review current Material Safety Data sheets for additional safety, handling, and disposal information.

Procedure

1. Obtain about 80 mL of distilled water in a 250-mL beaker and add 80 mL of methyl alcohol. There should be one liquid layer because methyl alcohol and water are miscible.

2. Weigh out approximately 55 g of potassium carbonate in the large weighing dish.

3. Gradually add the solid potassium carbonate to the methyl alcohol–water solution. Stir the mixture to dissolve the solid.

4. As potassium carbonate dissolves in the solution, two layers will begin to separate in the liquid. Continue adding potassium carbonate until the lower aqueous layer is saturated (undissolved solid remains).

5. Using a pipet, remove the upper liquid layer of methyl alcohol and add it to a clean 150-mL beaker. Repeat as necessary to remove all the methyl alcohol.

Demonstrations

Teacher Notes

6. Decant the aqueous potassium carbonate solution into a clean 150-mL beaker. (Do not transfer any solid.)

7. Add about 3 microspatulas of copper(II) sulfate pentahydrate crystals to the aqueous potassium carbonate solution. Stir to dissolve—the final solution should be clear and blue. (There may be some undissolved solid in the bottom of the beaker.)

8. Obtain about 80 mL of dry toluene in a 150-mL beaker. Add about 2 microspatulas of solid Sudan III dye and stir to dissolve. The solution should be clear and red.

9. Pour the three solutions into a large gas-collecting bottle in the following order:
   - Blue aqueous solution (contains K_2CO_3 and $CuSO_4$);
   - Colorless methyl alcohol solution;
   - Red toluene solution (contains Sudan III).

10. The result is a three-layer liquid! The bottom layer is blue (aqueous $CuSO_4$ and K_2CO_3), the middle layer is colorless (methyl alcohol), and the top layer is red (toluene and Sudan III).

Disposal

Please consult your current *Flinn Scientific Catalog/Reference Manual* for general guidelines and specific procedures governing the disposal of laboratory waste. The three-layered liquid may be saved in a sealed bottle. The top-most organic layer (toluene) may be poured into a shallow container and evaporated according to Flinn Suggested Disposal Method #18b. The remaining liquid layers may be disposed of down the drain with plenty of running water according to Flinn Suggested Disposal Method #26b.

Tips

- The three-layered liquid will keep for many months. Store in a sealed bottle and display in a secure location where the bottle will not accidentally be tipped over.

- Removing the methyl alcohol from the aqueous solution by pipet rather than simply decanting gives better separation of the aqueous and alcohol layers.

- Additional dyes may be added to achieve a red–yellow–green "traffic-light" density column. Add about one microspatula of methyl red indicator to the methyl alcohol layer after it has been salted out and separated from the aqueous solution. This will give a clear yellow solution. Add about one-half microspatula of potassium ferrocyanide $[K_4Fe(CN)_6]$ as needed to the blue aqueous layer to give it a nice green color. The result is red–yellow–green from top-to-bottom.

- The density of pure methyl alcohol (d = 0.791 g/mL) is less than that of toluene (d = 0.866 g/mL). In this three-layer density column, however, the toluene is the topmost (least dense) layer, because the methyl alcohol layer contains dissolved potassium carbonate, which increases the density of the solution.

- Replacing methyl alcohol with ethyl alcohol does not give a three-layer liquid, because the toluene and ethyl alcohol solutions are miscible (step 9). Ethyl alcohol can be used, however, to demonstrate the "salting out" effect with potassium carbonate (steps 1–4). Two layers will form when solid potassium carbonate is added to 50% ethyl alcohol.

Demonstrations

Teacher Notes

- The separation of 50 percent aqueous methyl alcohol in step 4 into two layers upon addition of potassium carbonate is due to the "salting out" effect. Adding an inorganic salt to water decreases the solubility of organic substances in the aqueous phase. Repeat this demonstration on a small scale with other inorganic salts to see if they have the same effect.

Discussion

Making a three-layer liquid demonstrates (1) the miscibility of methyl alcohol and water; (2) the "salting out" effect that results when a salt is added to aqueous methyl alcohol; (3) the immiscibility of toluene and methyl alcohol because toluene is a nonpolar solvent; and (4) the relative densities of the three solutions. Adding suitable dyes or indicators to the different liquid layers makes the phase boundaries clearly visible.

The ability of an organic compound to dissolve in water is drastically reduced when an inorganic salt (many different salts will work) is added to the water. This effect, called salting out, occurs because the water molecules bind strongly to the inorganic cations and anions in the salt solution, and thus are unavailable to solvate the organic compound. Salting out is used in two different ways to isolate and purify organic compounds from a reaction mixture. If the reaction mixture is an aqueous solution, salt is generally added to remove the organic product into a separate layer. Alternatively, an organic solution obtained by extraction with water may be "washed" with saturated sodium chloride solution to remove water dissolved in the organic solvent.

Safety and Disposal

Safety and Disposal Guidelines

Safety Guidelines

Teachers owe their students a duty of care to protect them from harm and to take reasonable precautions to prevent accidents from occurring. A teacher's duty of care includes the following:

- Supervising students in the classroom and laboratory at all times.
- Providing adequate instructions for students to perform the tasks required of them.
- Warning students of the possible dangers involved in performing the activity.
- Providing safe facilities and equipment for the performance of the activity.
- Maintaining laboratory equipment in proper working order.

Safety Contract

The first step in creating a safe laboratory environment is to develop a safety contract that describes the rules of the laboratory for your students. Before a student ever sets foot in a laboratory, the safety contract should be reviewed and then signed by the student and a parent or guardian. Please contact Flinn Scientific at 800-452-1261 or visit the Flinn Web site at www.flinnsci.com to request a free copy of the Flinn Scientific Safety Contract.

To fulfill your duty of care, observe the following guidelines:

1. **Be prepared.** Practice all experiments and demonstrations beforehand. Never perform a lab activity if you have not tested it, if you do not understand it, or if you do not have the resources to perform it safely.

2. **Set a good example.** The teacher is the most visible and important role model. Wear your safety goggles whenever you are working in the lab, even (or especially) when class is not in session. Students learn from your good example—whether you are preparing reagents, testing a procedure, or performing a demonstration.

3. **Maintain a safe lab environment.** Provide high-quality goggles that offer adequate protection and are comfortable to wear. Make sure there is proper safety equipment in the laboratory and that it is maintained in good working order. Inspect all safety equipment on a regular basis to ensure its readiness.

4. **Start with safety.** Incorporate safety into each laboratory exercise. Begin each lab period with a discussion of the properties of the chemicals or procedures used in the experiment and any special precautions—including goggle use—that must be observed. Pre-lab assignments are an ideal mechanism to ensure that students are prepared for lab and understand the safety precautions. Record all safety instructions in your lesson plan.

5. **Proper instruction.** Demonstrate new or unusual laboratory procedures before every activity. Instruct students on the safe way to handle chemicals, glassware, and equipment.

Safety and Disposal

6. **Supervision.** Never leave students unattended—always provide adequate supervision. Work with school administrators to make sure that class size does not exceed the capacity of the room or your ability to maintain a safe lab environment. Be prepared and alert to what students are doing so that you can prevent accidents before they happen.

7. **Understand your resources.** Know yourself, your students, and your resources. Use discretion in choosing experiments and demonstrations that match your background and fit within the knowledge and skill level of your students and the resources of your classroom. You are the best judge of what will work or not. Do not perform any activities that you feel are unsafe, that you are uncomfortable performing, or that you do not have the proper equipment for.

Safety Precautions

Specific safety precautions have been written for every experiment and demonstration in this book. The safety information describes the hazardous nature of each chemical and the specific precautions that must be followed to avoid exposure or accidents. The safety section also alerts you to potential dangers in the procedure or techniques. Regardless of what lab program you use, it is important to maintain a library of current Material Safety Data Sheets for all chemicals in your inventory. Please consult current MSDS for additional safety, handling, and disposal information.

Disposal Procedures

The disposal procedures included in this book are based on the Suggested Laboratory Chemical Disposal Procedures found in the *Flinn Scientific Catalog/Reference Manual*. The disposal procedures are only suggestions—do not use these procedures without first consulting with your local government regulatory officials.

Many of the experiments and demonstrations produce small volumes of aqueous solutions that can be flushed down the drain with excess water. Do not use this procedure if your drains empty into groundwater through a septic system or into a storm sewer. Local regulations may be more strict on drain disposal than the practices suggested in this book and in the *Flinn Scientific Catalog/Reference Manual*. You must determine what types of disposal procedures are permitted in your area—contact your local authorities.

Any suggested disposal method that includes "discard in the trash" requires your active attention and involvement. Make sure that the material is no longer reactive, is placed in a suitable container (plastic bag or bottle), and is in accordance with local landfill regulations. Please do not inadvertently perform any extra "demonstrations" due to unpredictable chemical reactions occurring in your trash can. Think before you throw!

Finally, please read all the narratives before you attempt any Suggested Laboratory Chemical Disposal Procedure found in your current *Flinn Scientific Catalog/Reference Manual*.

Flinn Scientific is your most trusted and reliable source of reference, safety, and disposal information for all chemicals used in the high school science lab. To request a complimentary copy of the most recent *Flinn Scientific Catalog/Reference Manual,* call us at 800-452-1261 or visit our Web site at www.flinnsci.com.

National Science Education Standards

Experiments and Demonstrations

Content Standards	Models of Organic Compounds	Making Soap	Preparation of Esters	Synthesis of Aspirin	Steam Distillation of Cinnamon	Cleaning with Charcoal	The Carbon Soufflé	Feeling Blue	Kaleidoscope Optical Activity	Salt-Out the Red, White, and Blue
Unifying Concepts and Processes										
Systems, order, and organization	✓									
Evidence, models, and explanation	✓								✓	
Constancy, change, and measurement		✓	✓	✓						
Evolution and equilibrium										
Form and function										
Science as Inquiry										
Identify questions and concepts that guide scientific investigation										
Design and conduct scientific investigations										
Use technology and mathematics to improve scientific investigations	✓		✓							
Formulate and revise scientific explanations and models using logic and evidence										
Recognize and analyze alternative explanations and models										
Communicate and defend a scientific argument										
Understand scientific inquiry			✓							
Physical Science										
Structure of atoms										
Structure and properties of matter	✓	✓	✓	✓	✓	✓			✓	✓
Chemical reactions		✓	✓	✓			✓	✓		
Motions and forces										
Conservation of energy and the increase in disorder										
Interactions of energy and matter										

National Science Education Standards

Experiments and Demonstrations

Content Standards *(continued)*

	Models of Organic Compounds	Making Soap	Preparation of Esters	Synthesis of Aspirin	Steam Distillation of Cinnamon	Cleaning with Charcoal	The Carbon Soufflé	Feeling Blue	Kaleidoscope Optical Activity	Salt-Out the Red, White, and Blue
Science and Technology										
Identify a problem or design an opportunity										
Propose designs and choose between alternative solutions										
Implement a proposed solution										
Evaluate the solution and its consequences										
Communicate the problem, process, and solution										
Understand science and technology		✓		✓						
Science in Personal and Social Perspectives										
Personal and community health										
Population growth										
Natural resources						✓				
Environmental quality										
Natural and human-induced hazards										
Science and technology in local, national, and global challenges										
History and Nature of Science										
Science as a human endeavor		✓		✓						
Nature of scientific knowledge		✓	✓	✓	✓					
Historical perspectives		✓		✓	✓					

Master Materials Guide

(for a class of 30 students working in pairs)

Experiments and Demonstrations

Chemicals	Flinn Scientific Catalog No.	Models of Organic Compounds	Making Soap	Preparation of Esters	Synthesis of Aspirin	Steam Distillation of Cinnamon	Cleaning with Charcoal	The Carbon Soufflé	Feeling Blue	Kaleidoscope Optical Activity	Salt-Out the Red, White, and Blue
Acetic acid, glacial	A0177			60 mL							
Acetic anhydride	A0156				15 mL						
Aspirin	A0201				1						
Benzoic acid	B0197			30 g							
Boiling stones	B0136				15						
Calcium chloride, dihydrate	C0196	8 g									
Charcoal, activated	C0202						3 g				
Cinnamaldehyde	C0400					10 mL					
Copper(II) sulfate, pentahydrate	C0102										1 g
Corn syrup	C0091									450 mL	
Dextrose	D0002								1 g		
Ethyl alcohol	E0007			60 mL	56 mL						
Hexanes	H0002					225 mL					
Iron(III) chloride, hexahydrate	F0006	14 g									
Iron(III) chloride solution, 0.1 M	F0045			50 mL							
Isopentyl (Isoamyl) alcohol	I0031			60 mL							
Magnesium chloride, dihydrate	M0146	11 g									
Methyl alcohol	M0054			60 mL							80 mL
Methylene blue solution, 1%	M0074								1 mL		
1-Octanol (Octyl alcohol)	Q0024			60 mL							
Olive oil	Q0004		45 g								
Potassium carbonate, anhydrous	P0038										55 g
Potassium hydroxide	P0058									6 g	
Propionic acid	P0187			60 mL							
n-Propyl alcohol	P0028			60 mL							
Salicylic acid	S0341			30 g	8 g						
Schiff reagent	S0180					30 mL					
Sodium bicarbonate solution, saturated	S0267			300 mL							
Sodium carbonate, anhydrous	S0051									25 g	
Sodium chloride	S0061					150 g					
Sodium hydroxide solution, 6 M	S0242		100 mL								
Sucrose	S0134								60 g		
Sudan III	S0152										1 g

Continued on next page

(for a class of 30 students working in pairs)   **Experiments and Demonstrations**

	Flinn Scientific Catalog No.	Models of Organic Compounds	Making Soap	Preparation of Esters	Synthesis of Aspirin	Steam Distillation of Cinnamon	Cleaning with Charcoal	The Carbon Soufflé	Feeling Blue	Kaleidoscope Optical Activity	Salt-Out the Red, White, and Blue
Chemicals, continued											
Sulfuric acid, 18 M	S0228			30 mL	2 mL			60 mL			
Tallow	T0084		105 g								
Toluene	T0019										80 mL
Glassware											
Beakers											
50-mL	GP1005	15		15							
150-mL	GP1015										3
250-mL	GP1020	15		30		1					1
400-mL	GP1025		5		15						
400-mL, tall-form, Pyrex®	GP1059					1					
500-mL, tall-form, Pyrex®	GP1060								1		
Bottle, French square, 240-mL	AP8452										1
Capillary tubes, one end open	GP7047			1 pkg.							
Erlenmeyer flasks											
50-mL	GP3021			15	30						
125-mL	GP3040			15		1					
250-mL	GP3045			15							
500-mL	GP3050								1		
Funnel	GP5050			15		1					
Graduated cylinders											
10-mL	GP2005	15	30	15							
100-mL	GP2020								1	1	
Organic distillation set*	AP6351				15						
Pipets, Pasteur	GP7042			15	15						
Pipets, Volumetric, 25-mL	GP7031										1
Separatory funnels, 125-mL	GP5059				15						
Stirring rods	GP5075	15		15	15		1	1			1
Test tubes, 13 × 100 mm	GP6063	75		45	30						
Test tubes, 16 × 150 mm	GP6066		120								
Test tubes, 25 × 50 mm	GP6069			15							
Watch glasses	GP8006		60	15							

*Organic distillation set includes a one-piece custom distillation head (3-way adapter, condenser, and outlet adapter), as well as both a round-bottom distilling flask (250-mL) and a round-bottom receiving flask (100-mL).

Continued on next page

Master Materials Guide

(for a class of 30 students working in pairs)   **Experiments and Demonstrations**

	Flinn Scientific Catalog No.	Models of Organic Compounds	Making Soap	Preparation of Esters	Synthesis of Aspirin	Steam Distillation of Cinnamon	Cleaning with Charcoal	The Carbon Soufflé	Feeling Blue	Kaleidoscope Optical Activity	Salt-Out the Red, White, and Blue
General Equipment and Miscellaneous											
Balance, centigram (0.01-g precision)	OB2059		3	3	3	3	1	1	1		1
Clamp, buret, plain jaw	AP8354				15	30					
Filter paper, 12.5 cm diameter	AP8998				15		1				
Hot plate	AP4674		3	5	5	5					
Heating mantle	AP4372					15					
Melting point apparatus	AP6042				1						
Microspatula	AP9015										2
Organic small-group model set	AP5453	10									
pH test paper, pH 0–13	AP5894		1								
Pipets, Beral-type, graduated	AP1721		60	60	60	45			1		
Pipet filler	AP1887										1
Polarizing film	AP8913									3	
Ring support, 2-inch	AP1320				15		1				
Rubber stopper, size 7	AP2229								1		
Rubber tubing, 1/4 inch	AP1176					30					
Scissors	AP5394								1		
Spatula	AP8338		15		15						
Support stand	AP8226				15	30	1				
Test tube clamp	AP8217			15							
Test tube rack	AP1319		15	15							
Thermometer, digital	AP6049		15	15	15	15*					
Thermometer holder, rubber	AP1636					15					
Tongs, utility	AP1359							1			
Variable transformer (temperature controller)	AP4434					15					
Wash bottle	AP1668		15	15	15				1		
Water, distilled or deionized	W0007, W0001		✓	✓	✓	✓			✓		✓
Wax marking pencils	AP8291			15	15						
Weighing dishes, small	AP1277		30				1				
Weighing dishes, large	AP1279					15					1

*Regular glass-bulb thermometers are needed for steam distillation.